Inventing the 20th Century

100 Inventions that Shaped the World

No. 765,975.

PATENTED JULY 26, 1904.

E. P. HOOLEY.

APPARATUS FOR THE PREPARATION OF TAR MACADAM.

APPLICATION FILED NOV. 3, 1902.

NO MODEL.

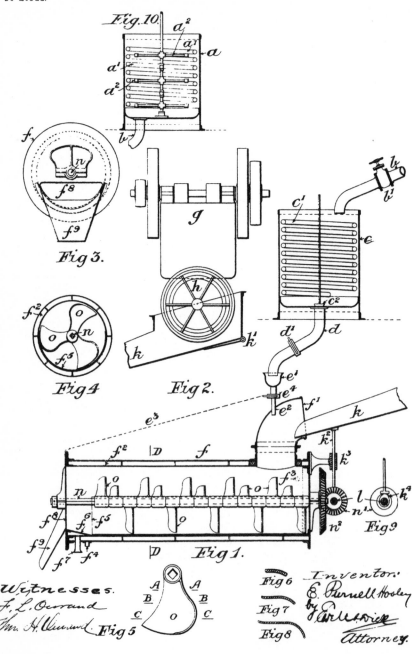

Fig. 10.

Fig 3.

Fig 4.

Fig 2.

Fig 1.

Fig 9.

Witnesses.
F. L. Ourand
Wm. H. Ourand.

Fig 5.

Fig 6

Fig 7

Fig 8

Inventor:
E. Purnell Hooley
by
Attorney.

Inventing the 20th Century

100 Inventions that Shaped the World

From the Airplane to the Zipper

Stephen van Dulken

With an Introduction by Andrew Phillips

NEW YORK UNIVERSITY PRESS
Washington Square, New York

First published in the U. S. A. in paperback in 2002 by
New York University Press
Washington Square
New York NY 10003

Library of Congress Cataloging-in-Publication Data
van Dulken, Stephen, 1952–
Inventing the 20th century: 100 inventions that shaped the
world: from the airplane to the zipper / Stephen van Dulken.
p. cm.
Includes bibliographical references and index.
ISBN 0-8147-8808-4 (cloth : alk. paper)
ISBN 0-8147-8812-2 (pbk.)
1. Inventions—History—20th century.   I. Title.
T20.V36   2002
609'.04—dc21   00-041141

Designed by Bob Elliott
Typeset by Hope Services (Abingdon) Ltd.
Printed in England by St Edmundsbury Press,
Bury St Edmunds

CONTENTS

AUTHOR'S PREFACE

BY STEPHEN VAN DULKEN

THIS book describes 100 significant, or at least interesting, inventions from the 20th century. They have been chosen for their stories: how the inventions were thought of, how the problems of making the product were solved, or how they were marketed. Each individual account is accompanied by a drawing or relevant text from the patent specification. Some were entirely new products, while many were 'improvements'. Many were important in industry, saved time, or in other ways made life more convenient. A few at least saved lives, while others were simply fun. They also provided employment for many.

The process of invention is a mysterious one and not one that can be reduced to a formula. Genius is at least in part the ability to think 'divergently', where the mind works differently from that of most people. But hard work is necessary as well. Thomas Edison was right when he said that genius was 'one per cent inspiration and ninety nine per cent perspiration'. Few inventions spring fully formed from anyone's mind. In most cases, although there may be a spark of an idea, the way of making it into a useful product still has to be worked out in detail. This may be necessary because the product simply would not sell unless it were cheap to make, or easy to use. It also has to be continually developed to make it cheaper or more useful, or to create other uses, in order to keep ahead of envious competitors.

Edison did not mention one element that comes into many of these stories: luck. Many ideas come to inventors by chance, perhaps as a result of an observation, or because they happened to be the right person to be asked about a work problem, or to overhear such a conversation. Others may have seen or heard something, but only a few actually did anything about it. They then had to work hard to make sure that the idea worked in practice, often over a period of years. In other cases an inventor was fortunate enough to have a friend or relative who was well placed to manufacture or to encourage the sale of the product. Bad luck was, perhaps, lacking such a friend or relative, and certainly money, since there is little point in inventing and patenting a product if no one knows about it, or cannot buy it from anyone. Ralph Waldo Emerson was not quite right when he said that a man who made a better mousetrap would have the world beating a path to his door.

It is astonishing how many of these stories involve a product being rejected by reputable companies as not being worth marketing, or where the original use of the product had to be dropped in favour of the 'obvious' use. What to us may seem obvious is not so obvious before it has been thought of. It is strange how long it sometimes takes for an 'obvious' idea to be carried out. Hindsight is a wonderful thing in the world of invention as in the stockmarket, current events and much else. Sometimes, of course, the product is not possible or at least not worthwhile until a scientific discovery is made, or a new material is available for use. The marketing of the products comes out in some of the stories as well. This too can involve highly creative thinking.

I hope that this book will provide hours of amusement and gentle education for its readers. Those who wish to explore these, or their own ideas, further should refer to the *Further Reading* section at the end of the book. This book is dedicated to all of you who feel a deep-down urge to create or design something better, even if you are not sure yet what it is. Robert Browning said, 'A man's reach should exceed his grasp'. Reach for your star.

In brief: something about the patent system

The patent system has evolved over a long time and is by no means consistent between countries. The US system, for example, is quite different from most of the rest of the world. A broad-brush, simplified version, taking in elements reflected in this book, is given below.

Patents are granted protection by applying to the patent office of each country where protection is wanted. It is normal but not required in many countries to apply first to the patent applicant's home country (in Britain it is mandatory to do so). By international agreement (the Paris Convention of 1883), foreign protection can be secured by applying elsewhere within 12 months of the 'priority date', and claiming that original date. Where necessary the foreign applications are made (and subsequently published) in that country's language. The point of this is that an invention has to be new in order to be accepted by each country, which separately decides on its validity without reference to each other. The 'novelty' of the invention is determined by excluding patent applications, which are similar to those filed before the priority date, or articles or books published before the priority date. These are all called 'prior art'. The only exceptions to this rule are the USA and the Philippines, where the important date is not the priority filing but rather the date of invention, which is rather more difficult to prove. Detailed, witnessed laboratory notebooks are kept to provide such evidence.

Each application as published will contain a description of how the invention works, with US patents often giving extra comments on prior art; drawings, if needed to show how the invention works; and claims which outline the monopoly on the new idea that is requested (in any initial published stage) or granted. Sometimes a search report listing prior art is attached at the back, or at the front (US patents). The British system worked as follows. Until 1978 the patent application was formerly published as 'accepted' and later 'sealed' as a granted, valid patent. Since then there are two stages, coded A for applications, which are published 18 months from the priority date, and B for the patent grant. The publications were numbered within the year of application until 1915, hence GB 17433/1901. From 1916 they were published in a series beginning with 100,001, which changed to a series beginning with 2,000,001 when the two stage-publications began.

Two international schemes also began in 1978. The Patent Cooperation Treaty provided for an application to be filed at the World Intellectual Property Organization which would publish it 18 months from the priority date. These applications then have to be carried on separately in the countries in which the applicant is interested. This scheme has become very popular in recent years among large companies. Its publications are numbered within the year of publication so that in 1990 there were numbers like WO 90/10935. The other international scheme is the European Patent Convention. Again it publishes an application 18 months from the priority date but this time it will publish a granted stage as well. Its publications were numbered from 1.

The US system publishes a single, granted stage. They are numbered in a sequence beginning with 1 in 1836 which has recently passed the 6 million mark. By a 1999 law this scene will shortly become much more complicated as a partial two-stage policy is introduced.

Many countries use the A and B codes mentioned for the British two stages although there are exceptions. The USA does not at present use such codes.

By international agreement, each country or patent authority has a two-letter code, which is used in this book to indicate their patent publications. Those used in this book are CA, Canada; CH, Switzerland; DE, Germany; DK, Denmark; EP, European Patent Convention; FR, France; GB, Great Britain; HU, Hungary; JP, Japan; US, USA; WO, Patent Cooperation Treaty. For those who wish to research the old British patents this author has also written *British patents of invention, 1617–1977: a guide for researchers*, published by the British Library in 1999. There is no equivalent publication for US patents.

ACKNOWLEDGEMENTS

THIS book grew out of an exhibition of illustrations of 20th-century patents put together by David Townsend in the former home of the British Library's science, technology and business wing at 25 Southampton Buildings, London in 1998. I should also like to thank Anthony Warshaw and David Way of the British Library Publishing Office, and Bob Elliott who designed the book.

INTRODUCTION

BY ANDREW PHILLIPS

FROM computers to catseyes, from genetic fingerprinting to fish fingers, from microwave ovens to silicon chips, this book highlights a selection of inventions and discoveries from the 20th century—'the century of invention'—which have transformed how we live, think, work and play. A great chemist of the century, Linus Pauling, said of innovation, 'The way to have good ideas is to have lots of ideas and throw the bad ones away', while the inventor, Trevor Baylis, remarked, 'Invention isn't some impenetrable brand of magic; anyone can have a go'. Certainly no century has accelerated change, nor built on innovative successes, nor rectified failures as quickly as the 20th century did.

The 18th century has been described as the age of agriculture, the 19th that of manufacture, the 20th that of energy while the 21st will be the age of information. But the 20th century has also been called the century 'of science', 'of biology', even 'of America'. Like all generalities these contain some truths but each century builds on and overlaps the achievements of its predecessors. So, across the bridge of the years 1760–1830, Britain changed from a generally agricultural country into a largely industrial one via inventions such as Watt's improved steam engine and Hargreaves's spinning jenny. In similar fashion the 19th century's later decades saw Bell's telephone, Edison's light bulb, Marconi's radio, the motor cars of Daimler and Benz, Röntgen's X rays, Tesla's electric motor for home and work appliances. All of these influenced profoundly ways of living in the century which followed.

Discovery and invention: inventors and discoverers

The different labels of the 20th century are apt. The century 'of energy' charted discoveries concerning the structure of the atom through the work of such physicists as Albert Einstein and Ernest Rutherford. The century 'of biology' discovered the means of genetic identification of the living and dead, of grafting bodily limbs and transplanting organs, of reactivating and reproducing living organisms. The century 'of science' traced the first evidence of different galaxies and the 'Big Bang' theory to describe the universe's origin and evolution. And much more. The 20th century journeyed into space; split the atom; cloned a sheep, a calf, a monkey; developed the computer, radar and television; built jet aeroplanes, rockets, satellites, space stations and shuttles. The ever-quickening pace of these inventions, and the scientific and medical discoveries which set the stage for so many of them, was such that the century's later years oversaw the beginnings of an electronic and communications revolution which will outleap Gutenberg's invention of printing or the industrial revolution itself in its everyday impact on humankind.

Many of these inventions and discoveries, or transfers between technologies, are the products of great individual minds. But many were developed by a host of contributors (even committees: for example, the electronic protocols of the World Wide Web, though the Web or Internet also owes much to individuals such as Leonard Kleinrock) building on each others' advances so that origination has been long forgotten in the same way as most English-language speakers do not know which of its familiar phrases William Shakespeare or William Tyndale coined. Or, just as most of the myriad users of the hypertext linkage of the World Wide Web are not aware of its designer—Tim Berners-Lee. Some inventions sprang from single moments of insight. Trevor Baylis, whose inventive imagination was early fired by playing with the Meccano sets described in this book, conceived the idea of a clockwork radio (1991) when watching a television programme about Africa. But he went on to say that 'Until you have a working product on the market it's not over'.

If the majority of 20th-century inventions originate or were patented in a relatively small number of countries chiefly in Europe and North America this reflects the geopolitical balance of international power during the century and where the chief resources for scientific, medical—and military—development lay. This must not overlook, though, the gifted inventors of other parts of the world, still less the cases of those emigrés who left their country of birth in search of a better life or to escape political and religious persecution by the far too many totalitarian regimes which the 20th century spawned. The physicists Albert Einstein, Enrico Fermi, Hans Bethe, Edward Teller, the ballpoint pen inventor Laszlo Biro, the mathematician and computer designer John von Neumann and the iconoscope/ television inventor Vladimir Zworykin, are all examples of scientists or inventors who moved to the 'New World'.

Inventing involves creating something which wasn't there previously: machines, processes, instruments, materials—almost anything. Many individual inventions are impossible without preceding and pervasive scientific or medical revelations. The atomic bomb which ended World War II could not have been assembled without the conclusions in the century's first decades of scientists like Einstein, Rutherford and Niels Bohr whose research however was not guided in any way by motives of war and conquest. On the other hand therapy and technology concerning human genes, and the mapping of the 100,000 or so genes in the human body (the ongoing Human Genome Project), would not have been possible in recent decades without computing technology.

Men born in 1900 would soon become familiar with the Model T Ford (1908), the 'all struts and wings' aeroplanes of Orville and Wilbur Wright (1903) as well as begin to use safety razors (1901). Their homes would begin to contain those devices which so many people in the Western world have come to expect to be on hand as easily as they breathe the air: the electric washing machine (1909), the vacuum cleaner (1901), television (1923) and other inventions for 'everyday' which this book includes. Now, in 2000 and ensuing years, their grandchildren and great-grandchildren are travelling by supersonic aeroplane and shopping internationally for all kinds of goods over electronic networks using home computers: the Concorde airliner first flew in 1969, in the same year that computers in two institutions in California first 'talked' to each other and ushered in the age of the Internet. But tarmac® road surfacing (1902) is still used each day all over the world 100 years later.

Only a century's journey spans the telephone and the Internet's electronic mail. But what is the colouring on the background canvas of the painting which illustrates such a march of discovery, and which stands behind some of the detailed inventions with which this book deals? Very many colours of new knowledge lie at the back of the canvas of the 'century of science'. If there is one scientist whose artistry touches many parts of this canvas it is Einstein with his recasting of our comprehension of space, time and the universe. His influence in altering these concepts ranks with Newton's and Galileo's.

Advances of nuclear and atomic physics including the theory of relativity and quantum theory, especially during the century's first 40 years, led to so many consequences for space travel, hydrogen weaponry, semiconductor and laser technology, even analysis of molecular behaviour. Bohr's 'quantum leaps', Planck's 'quanta', Gell-Mann's (and James Joyce's) 'quarks', Heisenberg's 'uncertainty principle' entered the common language as well as expressing a revolution in physics. Sakharov, Oppenheimer and Hahn meanwhile became almost as famous for their fears concerning the destructive power of nuclear technology, if wrongly used, as for their innovative contributions to the science.

The human mind's complexity was examined as never before by the studies in psychoanalysis and psychology of Freud, Jung and Piaget introducing for us '. . . a whole climate of opinion/under whom we conduct our different lives'. Pathfinding steps to transform mathematics and economics, respectively, were cut by Gödel's 'incompleteness theorem' and Keynes' 'general theory'. In molecular biology Crick and Watson revealed in 1953 the double-helix structure of DNA which was to show how hereditary data of life can be traced and to activate a new age of genetic research and engineering. The work of Hubble in astronomy proved that the universe is expanding (confirming Einstein's theory) and that other galaxies exist in space beyond the Milky Way. Lemaitre developed the Big Bang explanation of the universe's origin and Hawking's mathematical approach advanced greatly the theory and research of 'black holes', where massive stars have shrunk and disappeared.

If understanding the contours of the universe were changed by these discoveries immense benefit was brought to living in the world by achievements in bacteriology and virology. Fleming's discovery of penicillin and its properties in the 1920s would lead on to life-saving drugs to combat bacterial infection and to overcome most instances of diseases like tuberculosis and syphilis. Florey's and Chain's contribution after 1939 to the effective production of penicillin, to save countless lives in World War II and later, ranks with Fleming's discovery. The vaccines of Salk and Sabin in the 1950s defeated the worst ravages of poliomyelitis while the AIDS virus and HIV infection were identified at last by Gallo and Montagnier in the early 1980s. This was an essential first step towards rolling back the mutating HIV virus which many have come to view as the 20th century's fatal equivalent of the 14th century's Black Death. These areas of research together with mention of some of the outstanding practitioners associated with them are indivisible from the momentum of so much invention across the 20th century and bear witness to the tide of its innovation.

It is beyond this book's scope to paint in all the colours of this picture of discovery. What can be seen among the century's inventions, which form individual images in the picture's foreground as it were, is the accelerating progress of *clusters* of inventions and of the machines, processes and instruments of all kinds which have emerged within them. Some foremost examples of these are glimpsed from the following—very broad and loosely defined—'family groups' of inventions which, with extensive cross-breeding, were born and multiplied during the 20th century.

Air, sea and space technology

Whittle patented the air jet engine in 1930 and von Ohain built the first jet aeroplane for Heinkel 9 years later. In 1939 Sikorsky designed what was to prove the standard for the modern helicopter. But if 9 years were to separate the jet engine's patent from the first flight, only 30 years were to span the Heinkel He178 and the flight of the supersonic Concorde airliner. Nuclear submarines and ships had to await the 1950s for their first appearance but Oberth's and von Braun's extensive rocket technology in Germany of the early 1940s already owed much to Goddard's pioneering launches in America during the previous two decades. All acknowledged their debt to Tsiolkovsky's proposal in Russia as early as 1903 that multi-stage rockets could be used in space. From 1960 onwards von Braun and American and Soviet scientists were to achieve the rapid invasion of space and travel to the Moon via Sputnik satellites, Saturn rockets, Apollo spaceships, Columbia space shuttles, the Mir and Salyut space stations and in the 1990s the concept of an international space station.

Synthetics and substitutes

The first half of the 20th century revolutionised the technology of synthetics with major products evolving within ten-year steps. Baekeland's plastic (1909) was followed by cellophane in the century's second decade, by synthetic rubber in the third, by perspex, polythene, teflon, Carothers' nylon in the fourth and bubble-wrap and commercially produced PVC during the years of World War II. These materials would transform what people used and wore everyday at home and at work, in the richest and in the poorest countries.

Radio wave and light wave technology

The development of radar for military means and air defence from the mid-1930s, due to the work of Watson-Watt and others, not only had remarkable bearing on the outcome of World War II but also led to many applications for ground, sea and air in times of peace and to encourage spin-off technology. Maiman's invention in 1960 of the laser, to generate laser light beams, bore increasing implications for science and industry (for example, drilling and metal working) and opened a new dimension to medicine and intricate surgery.

Computers

Modern information technology's foundations were laid in the 1940s by Turing, Eckert and Mauchly together with the development of essential components such as Eisler's circuit board and the transistor invented by Shockley and his colleagues. The microchip/silicon chip invented at the close of the 1950s by Noyce and Kilby and Hoff's microprocessor ('computer on a chip') of 10 years later set in place the miniaturisation which stimulated the explosive development of personal computers and their systems from the 1970s to the present. In this extraordinary journey of technical progress many ordinary home computers are more powerful than the mainframe ones used by the NASA space agency only 30 or so years ago.

Electronic communications

In the 1920s John Logie Baird, Zworykin and Farnsworth all shared in different ways as inventors of television, and the transmission of electronic images, which has so influenced much of the world since the advent of mass television in the 1950s. The visual screen united television and the first home computer generation of the 1970s. The Internet has now delivered computer access to text, sound and vision from a host of sources with a speed of development and response which most people, 30 years ago, would have thought unimaginable in their lifetimes.

In 1991 Tim Berners-Lee's hypertext language and links were spun into an overall pattern which he introduced as the World Wide Web, bringing direction to often unmapped 'cyberspace' and pathways and charts to the 'howling wastes of the Internet'. The influence of his network design, and other improvements in information technology generally, can be shown by the number of Internet users and their rate of increase. There were two users in 1969, 100 in 1977 when Al Gore introduced the phrase 'information superhighway', 1000 in 1984, more than 2 million in 1991 with the launch of the World Wide Web, over 60 million in 1996 and more than 250 million at the beginning of 2000.

Embryology and genetic engineering

The stimulus of the discovery of DNA, the chemical pattern of instructions which controls how organisms work and look, helped a revolution in genetics and

5

medicine and in people's lives and hopes. Birth control pills, developed in the 1950s through the work of Pincus, Chang and others, removed much anxiety from countless families and relationships but also encouraged a sexual freedom which only the spread of the AIDS virus has discouraged—at least in many Western societies. In 1978 the first test-tube fertilisation baby was born in Britain and less than 20 years later Wilmut and his team cloned 'Dolly', a sheep produced from an embryo from a single cell of another adult sheep. Dolly later gave birth naturally: an event which may have an immense effect on the future prospects for the birth and life of all animals.

Medical diagnostic and surgical systems

The first successful kidney transplant was performed at exactly the half-way point of the century and new techniques of surgery advanced greatly in the later decades especially after the initial human heart transplants of the late 1960s by Barnard in South Africa. Medical technology saw consistent progress through the 20th century, so that now even pacemakers for the brain are being developed to treat illnesses such as Parkinson's disease, and important inventions arose in a number of countries.

Einthoven in the Netherlands designed an effective electrocardiograph as early as the century's beginning and it was there, too, that Kolff produced a kidney machine in the 1940s. A heart pacemaker was first fitted by Senning in Sweden in 1958. In the USA Drinker developed the iron-lung at the end of the 1920s and Gibbon accomplished open-heart surgery employing a heart-lung machine in the 1950s. Britain made particular contributions between the 1950s and 1970s to medical scanning systems. Drawing upon sound wave technology (like sonar systems in submarines which detect other ships) and upon computing and X-rays, Hounsfield and Donald and others developed ultrasound, brain and body scanners to build up and reassemble internal pictures of the body and its parts.

Much of this invention has flourished through teams of researchers and a good number of invention dates crystallise too simply extensive periods of research which preceded them since dates of inventions and scientific breakthrough can often be quoted variously: for example, the date of first inspiration, or the date of patenting or the date when an invention was unveiled to the world's eye. The fast-flowing developments of the 1980s and 1990s, in particular, are almost too numerous to isolate singular dates.

The rolls of honour represented by these inventors (and many more), some to be enduringly famous some never to be so, have left a distinctive handprint on 20th-century living as well as on that of the century through whose door we have now entered. But Baylis's earlier comment about 'working products' is a reminder that successful inventing means meeting human, social and economic needs as well as often generating financial benefit. It is noticeable how many inventions, which this book cites, have alleviated repetitive—often back-breaking and mind-numbing—work which could once wear down human spirits and make the young old before their time; or, have added to enjoyment of greater leisure which the century's inventions created for legions of people.

History, economics, politics: the hour-glass of the 20th century

What is the historical and economic shape of the century which demanded and supported so much invention? No previous century has been affected by such major wars, particularly two World Wars (1914-18 and 1939-45) in which more than 60 million people lost their lives. A century, which eradicated smallpox disease and hugely reduced the threat of poliomyelitis, could also contrive man-made slaughter and destruction on a scale unimagined previously—often through technological and industrialised methods pioneered by the century's brilliant inventions and prowess in mass-production—and bring about too the scientific means to destroy our planet Earth.

In many ways the 20th century's historical shape seems curiously like an hour-glass, the narrow opening between the two vessels being World War II which so influenced science, technology, industry and much of the political pattern of nations for the second half-century. The hour-glass's sands have slipped through to meet the glass's symmetry of time and some of the overall patterns in each vessel match or resemble each other. But most of the grains of sand have shifted their positions completely. Economic, political and military developments have been major causes of these shifts.

The century began in the first era of global markets from 1870 to 1914 and imperial domains where the Great Powers competed in terms of trade and industrial size, determination to annex colonial territories and capture economic spheres of influence, or exerted their international presence by sheer population mass: the USA; Western Europe, especially industrial Britain and Germany; a modernised Japan; and Russia. The Great Powers were uneasily at peace with each other while ferocious small wars occurred around them or entrapped them in conflicts in which the major power, though bloodied, would prevail: the Balkan wars of 1912 and 1913 and the earlier war between Britain and the Boer Republic are different examples.

Sixty years before, Alexis de Tocqueville had said of America and Russia that 'Their starting point is different and their courses are not the same; yet each seems to be marked out by the will of Heaven to sway the destinies of half the globe'. For many years after 1945, in the hour-glass's second vessel, this prophecy was fulfilled. Now at the 20th century's close and the new century's beginning the USA, having also made decisive interventions in both world wars, remains the predominant world power though the days of geographic empires have long gone. This can be seen in the scope of science and technology in the USA, international presence, military potential and—witness the longest economic boom of the 1990s in its recent history and a high-technology 'gold rush'—economic strength. Its influence helped ensure, too, that English became the *lingua franca* of the 20th century while after World War II the US dollar became the chief currency in international trading. As at the start of the century, so at its end the Great Powers have been at peace for 50 years: but bloodshed in countries such as Kosovo, Rwanda, Burundi, Afghanistan, has rivalled the savagery of the century's early wars and of massacres like those in Armenia (1909, 1915), while Russia has been steeped in a vicious war with Chechnya (with echoes of the British-Boer war).

Broadly, as in most of the 20th century, the USA is still now followed by the major countries of Western Europe in their new European Union federation but the emerging might of the world's most populous nation, China, may be expected to become a large factor in the present century adding to the technological progress of the Pacific Rim and especially of Japan, South Korea, Singapore, Taiwan, and Malaysia. Russia presently experiences some of the great economic and nationalist difficulties which it knew in a different guise in 1900, having travelled an immense journey of achievement in between. Another form of global economy now holds sway with multinational industries independent of countries' national borders, and with swiftly changing economic markets.

The 'tendons and blood vessels' of globalisation contained within the Internet and satellite technology are introducing an unprecedented speed of change. They also provide huge market scope for international interests while allowing the de-industrialisation of many of their workforce, re-creating in new form the domestic working of much earlier times. The present globalisation, like its early 20th-century predecessor, also shows great contrast: high-technology services for global office work foregather in places such as India's Bangalore alongside some of the world's most crushing poverty. Many may consider therefore that Karl Marx's insight, 'The need of a constantly expanding market for its products chases the bourgeoisie over the whole surface of the globe', is as pertinent as ever. Unlike the great conflicts of the 20th century (despite their many underlying economic reasons) the wars of the 21st century are more likely to be trade wars: to sell the most things rather than to kill the most people, as population continues to expand and our planet and its natural resources shrink further. The national empires of 1900 have been succeeded by the multinational commercial empires of 2000.

Many looked to communism (and some to fascism) replacing competition or economic differences between classes of people with social planning and economic equality among citizens. Neither fascist aggression for more 'living space' in the 1930s, however, nor communist controls between 1920 and 1990 could effectively bring wealth or sufficient food and resources to the peoples dependent on them. They brought ultimate failure with huge loss of life in Europe and Asia especially. Restless and liberal capitalism earned so much of the 20th century's wealth and is generating it today in the era of electronic information: but its price includes greater inequalities between billionaire and pauper and an unspoken emphasis on the survival of the fittest. Marx forecast, again, some of the dependence on technology and the spread of vast corporations which we see currently.

In between the global ambitions of nations and this globalisation of multinationals, at the entrance and exit of the 20th century respectively, a combination of economic competition and political rivalry among autocratic governments in Austria-Hungary, Germany and Russia ignited World War I. Along with its carnage this destroyed ramshackle empires and created new countries in regions such as the Balkans and the Middle East. As nations sought to pull their way clear from 19th-century rule and economic conditions, and repair economies shattered by World War I (worsened too by the war's aftermath and by continuing resentment between victor and vanquished) so the centralising regimes of communism and fascism were established during this revolutionary phase in the Soviet Union (principally Russia), Germany, Italy, Japan and elsewhere. The long success of the capitalist market economy, especially that of the United States, was checked also by the Wall Street crash (1929) and the Great Depression of the 1930s.

Overcoming these difficulties lay in improved economic achievement, better industrialisation, more innovation. The economic crisis of the 1930s also encouraged states to put their own commercial interests first before those of internationalist trade. The totalitarian regimes of Joseph Stalin in Russia and Adolf Hitler in Germany (and decades later of Mao Zedong in China) inflicted untold sufferings on many of their own, and later other, peoples and can never be dissociated from such horrors as the Holocaust and the Gulag Archipelago. In terms of economic systems they saw, however, centralised planning and encouragement of science, technology and industry as their chief path forward and the interests of the state came before all other values. Their nationalistic motives (and those of successive Japanese governments in the 1920s and 1930s) were to build the pre-eminence of their countries— and in Stalin's case to draw Russia completely into the 20th century—through economic strategies and military strength and preparedness. Ironically, but not surprisingly, this was not so different in many ways from the ambitions of supremacy which the imperial families of the Great Powers had held a generation before. In both a similar and rather different fashion to Stalin's and however much they distorted Marx's economic philosophy, Mao Zedong also sought to modernise China after 1949. (Conversely, Pol Pot's genocide in Cambodia in the 1970s was a senseless attempt in Marx's name to reverse technical and industrial advance.)

War in 1939, at least in Nazi Germany's intention, thus became a necessary step towards the supremacy of its economic future and technological regeneration. And themes of economic betterment echoed each other in the totalitarian and free worlds, though the latter was free of infamous motives and sought to avoid the immediate human tragedies to which Stalin and Hitler paid little account. There were similarities in the overall aims of improved economy and employment in the Soviet Union's collectivisation programmes; in Germany's industrial re-arming; in Franklin D. Roosevelt's New Deal of the 1930s to rekindle employment in the USA after the Depression and to reconstruct federal and state economies and markets in order to bring back some of the growth in prosperity which North America had experienced during the previous third of a century. This rekindling of employment in the USA was only accomplished fully by rearmament (as in Germany) and war production after 1941. Similarly, for example, World War II made British industry much more efficient and rescued much of the derelict farmland and farming of 1930s' Britain.

The breakdown of civilisation which so many peoples in Europe and Asia suffered during World War II did not slow technological advancement: rather, the needs of war speeded it. After that war's defeat of fascism in its most virulent form new blocs of countries were united or shepherded together under the umbrella of the superpowers' domination, realising for a time de Tocqueville's vision of the dual might of America and Russia. Nations like Britain and France divested themselves of their colonial empires, the bonds of which had already been loosened by World War II and growing cries for countries' independence, thus shaking off the 19th century.

But intense scientific competition rather than co-operation among the superpowers largely held sway until the 1970s as the 'Iron Curtain' became a metaphor for the division between Eastern Europe and the Soviet Union and Western Europe and North America, and between communism and capitalism. The Berlin Wall (1961) came to manifest physically not only the division of Germany after 1945 but also to symbolise the Iron Curtain during a time when the Soviet Union crushed independent dissent in Hungary (1956) and in Czechoslovakia (1968). Cooperation *within* these power blocs however, especially in the West through frameworks like the European Common Market, reaped many social and political benefits and the 20 years following 1950 brought almost continuous economic expansion to the most industrialised nations.

A long interlude of balanced power and global peace after 1945 continued during the 'Cold War', and its gradual subsequent warming, through possession by the superpowers of nuclear weaponry which could lead to another Armageddon if ever let loose. In the 1960s, 1970s and 1980s it became clear too that even a superpower might not defeat a determined nation when challenged in conventional warfare, as the USA discovered in Vietnam and the Soviet Union in Afghanistan. However, in the late 1980s 'the deceptively monochrome surface of the Soviet Union shattered'. The political systems of Russia and its dependent states with their centralised controls began to self-destruct through internal social and economic pressures as well as, in Latvia and elsewhere, reviving nationalism. More and more money had to be spent on retaining military strength, space technology, and maintaining a rusting industrial infrastructure: less and less could be distributed for the people's needs. These needs were reinforced by expectations since communist states could increasingly see the wealth, goods and new products which were always present—however variously—in countries free from communist centralisation after World War II. Moreover countries like Hungary and Poland had also incurred significant debts since they had borrowed money substantially in the hope that their later industrial expansion could repay it.

Mikhail Gorbachev's reforming policies of 'restructuring' and 'openness' in the Soviet Union foresaw some of the great change which had to come. During the same period China, under Deng Xiaoping, began to welcome capitalist concepts and encourage consumer booms. The 'ten days that shook the world' of the Russian Revolution in the first vessel of the century's hour-glass were thus matched by 'ten years that shook the world' in the dismantling of the Soviet Union's empire during the century's last decade. Individual states were re-created, some of whose territories looked again rather as they had done in the later 19th century. Many ethnic, religious and regional groups re-emerged prominently as over 20 countries regained independence after 1989, the year which saw the symbolic destruction of the Berlin Wall: Germany itself was reunified in 1990 after nearly 50 years, though within significantly changed borders from those of 1939.

These 'new'—but also old—countries could then embrace forms of capitalism seen increasingly but not unreservedly as the global wealth creator: even if they did not possess, as Russia did not, enough of the infrastructure and tradition which have sustained the capitalist system's growth in North America and Western Europe. The prize of freedom of opportunity often brought with it the price of uncertainty and more poverty. But the eruption of independent nationalism also awoke old hatreds which communist political structures had forcibly kept in check: the world saw fighting or 'ethnic cleansing' in Bosnia, Azerbaijan and Chechnya as examples of this. Savagery was still present in some grains of sand in the hour-glass after lying dormant for a century.

The rate of change across the world has accelerated since the fall of communism, keeping pace with some of the electronic revolution's rapidity: trade barriers have fallen, financial investment flowed more freely, people are increasingly following work rather than work being brought—by national government or big business—to people. The present emphases on individuals' achievement and entrepreneurial skills are making invention even more important at the dawn of the 21st century.

Mass-production and technology for war and peace

In the 20th century many gained hold of privilege and advantages which were once only in the hands of the few. There are many streams and patterns in the hour-glass representing economic and industrial factors which changed the face of the century of invention. It is appropriate perhaps to look further at two of the more important ones.

First, the coming together of mass-markets and mass-production. This took root among all the workers and consumers of the developed countries. It may be viewed especially as the individual legacy of the motor industry innovator, Henry Ford I, of whom it was said that 'He was the first to create a mass-market as well as the means to satisfy it'. His employees put his mass-production principles into practice thus helping to forge a 'blue-collar middle class' with the means to buy Model T motor cars. In the Western world particularly this introduced a vast market for new devices, inventions, and expectations of what people could buy. Many of the products in this book are the result of this expansion of people's opportunity.

Second, as noted before, war leads to invention as well as to devastation. Poison gas and armoured tanks in World War I; radar and atom bombs in World War II; later missiles, neutron bombs, 'stealth' air technology, the 'Star Wars' unfinished defence system proposals of the 1980s have all been part and parcel of the discoveries of science and technology. They have spun off many benefits, too, for times of peace. Radar was instrumental to the development of transistors and microwave ovens. Codebreaking Germany's Enigma secret codes in 1939–45 hurried in the post-war era of the computer.

The V2 long-distance rocket bore a wholly destructive purpose in World War II but rocket engineers always foresaw their technology introducing exploration of space and the planets. Military and political competition during the subsequent space race between the Soviet Union and the USA was a major factor in the speed of achievement which enabled the Soviet Union to put the first person in space in 1961 and the USA to land the first person on the Moon only 8 years later. 'Trickle down' technology again from the Apollo space missions helped the development of, for example, polarised sunglasses and kidney dialysis machines.

The collaboration of the Manhattan Project, which developed the atom bomb to destroy the Japanese cities of Hiroshima and Nagasaki in 1945, pushed forward hugely the frontiers of science and the careers of some of its most eminent practitioners. A leading organiser of the Project, the analogue computer designer Vannevar Bush, came to be seen as one of the creators of the 'military-industrial (and university) complex' after World War II which has had an immense effect on subsequent inventions for peace and war, global trade and world-wide arms. Dwight D. Eisenhower warned against the power and influence of this complex but it has not been only a creation of the USA. The Soviet Union and other major powers had, and have, their equivalent military-industrial structures and in many countries they continue to be a considerable source of national employment and income.

Some ethical issues

Science is neutral. No bleaker example of this is provided by I. G. Farben. Its scientists made the important breakthrough to produce synthetic rubber in the 1920s; later, Farben patented Zyklon B gas which was used in the vile mass murder of concentration camp victims. The ethical issues of science and invention for war, of human and animal reproduction, or of genetic engineering are for the world as a whole and its governments to be much concerned with and there are many such issues which the 20th century has bequeathed its successor.

After the atom bomb's destructive use in 1945 Einstein reflected of his own contribution, 'If only I had known, I should have become a watchmaker'. He also remarked to Max Born in the context of quantum theory, 'God does not play dice'. But men do. And the ceaseless searching and unpredictability of their scientific and inventive discovery will test their consciences, beliefs and values even further in the 21st century. One medical forecaster has predicted no more cancer in 100 years' time. We must pray too that this invention and discovery can mean that the 21st century will not leave to the 22nd some of the legacy it itself received. Despite the progress which this book shows in the year 2000, 800 million people are starving around the world, almost 35 million people world-wide are affected by HIV/AIDS (a disease unknown 30 years ago), and there still emerge deadly new strains of virus such as Ebola.

Conclusion

De Tocqueville referred to the 'destinies of half the globe' and this very brief historical narrative has not touched upon events across so much of the globe in the 20th century: particularly in Africa, Asia, and South America. Many of the problems which beset the most deprived countries in these continents—for example, most of the countries with the 10 worst child mortality rates are in Africa—await the help which the new century's invention and discovery can bring to them.

They also await the global political will to bring about that aid. Mass-production of food, due to innovative agricultural and food technology, thrived in the most developed nations in the second half of the 20th century. But, in the new century, this production must be passed widely to those most in need and the advantages and anxieties concerning genetically modified foods are an important example of this debate and prospect.

The inventions highlighted in this book have benefited people of virtually every nation. Some have helped combat the despair of disease, poverty, excessive (even unendurable) labour. Other inventions—though less illustrated by this book—have contributed to the ravages of war. What comes forward so often, however, from the examples described here is the individuality and initiative which characterise so many inventors who helped change the world between 1900 and 1999.

When the philosopher, Isaiah Berlin, wrote, 'Injustice, poverty, slavery, ignorance—these may be cured by reform or revolution. But men do not live only by fighting evils. They live by positive goals, individual and collective, a vast variety of them, seldom predictable, at times incompatible', he was speaking for inventors and discoverers down the ages. In 1999 a candidate for the Presidency of the USA observed that just as the tractor produced in the first half of the 20th century added to man's muscle, so the microchip invented in the second half of the century supplemented man's brain; and forecast that soon in the 21st century, 'A supercomputer would fit on a speck of dust'.

And who would discredit such prophecy when the 20th century saw life imitating art in many different ways? The real achievements of science and technology fulfilled some of the imaginative forecasts of writers of science fiction or 'futures' such as H. G. Wells, Aldous Huxley, Robert A. Heinlein and Arthur C. Clarke. Clarke is considered to be the originator of the idea of satellite communication and was called even by the NASA space agency 'the godfather of the space race'.

In the light of this continuing explosion of scientific, technological and medical development, fuelled by electronic communications across the world, some are wondering whether the human mind will cope with the scale of change. Or cope with, say, the prospect of human cloning or electronic devices now being implanted in the nerves and muscles of the human body or new inventions being designed increasingly by computers rather than by those who made the computers.

But human feelings change less than the alterations we impose on our external world, even in the swiftest times. Philip Larkin, the poet, wrote memorably, 'What will survive of us is love'. A man and a woman both born in 1899 and who were married in 1925, after experiencing the advent of the motor car, computers and the moon landing—the whole picture of the 'century of invention'—could still tell a London newspaper in the 1990s, 'Nothing has been more important over the last century than our time together'.

Today and tomorrow

Mandell Creighton, Bishop of London, died a year into the 20th century. He had earlier written that the best of human endeavour should provide 'wisdom for the present and hope for the future'. Almost everyday a century later, newspapers, TV and the Internet report new inventions and developments in research. Experts in all fields at the end of the 20th century forecast scientific, technical, medical and other achievements which they foresaw being brought about in the 21st century. Many of these predictions will be fulfilled. However, as written in the Koran, 'Those who claim to forecast the future are lying even if they are later proved to be right'. Equally inventions will come to pass which are not foreseen or anticipated because even as people become more advanced, expert, knowledgeable, and specialised much of the universe remains a mystery some of which may be illuminated only slowly and step-by-step, as the Persian poet-astronomer Omar Khayyam wrote:

> There was a door to which I found no key
> There was a veil past which I could not see.

But two statements seem especially appropriate in the year 2000. Much progress at present concerns genetics, cloning, prolonging or helping life. In Britain it is now being recommended that cloning of human embryos should be permitted for medical research into heart, liver and kidney diseases: some see this as the beginning of human body cloning. In view of this it is interesting to recall that in 1973 the theoretical physicist, Werner Heisenberg, said 'If I was starting out today I wouldn't go into physics. I'd be a biologist. That is where all the exciting work will be done in the next several decades'.

The second statement comes in the first year of the 21st century from Kofi Annan, the Secretary-General of the United Nations, on the publication of the report *We the Peoples* . . . describing many of the needs and threats which we face:

. . . [Globalisation] has brought many benefits but it also has a negative aspect. Whole regions are excluded especially Africa. We see incredible inequalities and exclusion which, if unchecked, will create serious problems and lead to a backlash . . .
. . . We should protect future generations from the threat of war and today the threat is mostly from internal not international wars . . .
. . . [The immediate needs are] the alleviation of poverty and the fight against AIDS. Also, the possible use of technology to allow Third World countries to leapfrog many of the painful development changes that others had to go through. Last but not least, the maintenance of the environment . . .

Trevor Baylis's clockwork radio was conceived with its use in Africa especially in mind and he has spoken of the concept of a clockwork laptop computer to help those countries too. It is not only the world's political rulers and religious leaders but also its discoverers and inventors—free of cult or creed—who must sustain Annan's and Creighton's 'wisdom for the present and hope for the future'.

Note on the decades of the 20th century

In the pages that follow, the selected inventions of each decade are introduced by a very short account of some milestones of history, discovery and technical progress which the decade witnessed. My accounts are necessarily very selective. It has been said that we have four basic ways of understanding and dealing with the universe around us and of coping with the fear spoken by the words of A. E. Housman's poem:

> I, a stranger and afraid
> In a world I never made.

These are: man the builder and worker, imposing his will on the world and changing its landscape; man the discoverer and thinker, forever looking to new horizons of knowledge; man the artist, changing perceptions of everything around us; man the believer, coming to terms with his existence through belief in the ever present influence of a deity.

My brief decade-by-decade introductions touch only on some achievements within these first two ways of viewing human circumstances. They do not include, for example, references to 20th-century arts and culture and many momentous events and famous names of the century's history are not mentioned. Rather they seek to refer to some historical trends and milestones which provide some context for the progress of invention and discovery through the century, and to cite individual discoveries which may not have been mentioned, or alluded to, elsewhere in the book. Again, most of the references concern events in Europe and North America, the most influential continents in overall political and economic terms during the century and where the majority of inventions and discoveries were produced and progressed: quickening as the decades passed.

1900-1909

IT was the high tide of the imperial age, with expansionist empires and increasing industrial capacity. A fifth of the world's land was occupied by the British Empire embracing 400 million people, three-quarters of them in India. Britain had long been called 'the workshop of the world' but by the end of the decade and the start of the next Germany had overtaken Britain in output of coal and steel and the USA was advancing to outstrip all. Though the major regimes of Europe appeared outwardly stable there was much internal unrest, especially rioting and mutiny in Russia in 1905. There were acute tensions between Austria-Hungary and the Balkan states and ethnic massacres in the Turkish Empire. The Young Turks under Enver Pasha seized control in Turkey in an attempt to bring the Ottoman Empire into modern times though this would only become effective under Atatürk decades later. Unemployment encouraged the emergence of a vigorous labour movement in Britain. But there David Lloyd George's 'people's budget' and the introduction of old age pensions and health/unemployment insurance were to prove landmarks towards a more equal society down the century.

Beneath the seemingly endless vista of their power and strength much was beginning to undermine Europe's ruling houses. Revolutionary fervour stirred in Russia because of the conditions of workers and peasants, the crew of the battleship *Potemkin* mutinied and there was bloodshed on the streets of St Petersburg and Moscow. Tsar Nicholas II allowed a parliament, the Duma, to have limited powers but this was dissolved when it clashed with the policy of the Tsar and his statesmen. The road to Lenin's revolution began to be laid. The unhappiness of the Russians was inflamed by humiliation in war against Japan in Manchuria. Japan seized Port Arthur and sank Russia's Baltic fleet, which had sailed around the world, in the battle of the Tsushima Straits. The USA's Theodore Roosevelt won a Nobel prize for brokering Russo-Japanese peace.

Germany's armaments programme gathered pace, particularly in building a navy to compete with Britain's, with the manufacturer Krupp of Essen to the fore. The first modern submarine appeared and Britain launched the *Dreadnought* battleship. Modern Japan emerged after its success in war against Russia. Emigration from Europe to the USA was at its peak and there was enormous commercial and industrial growth in the USA, especially in steel, railways, oil and banking. An illustration of this prosperity was Carnegie's sale of his company to US Steel for the huge price of $447 million.

The ambitions of the great empires did not go unchecked. European powers had long exploited the weakness of China but the Chinese taking part in the Boxer rebellion in Beijing, encouraged by the Dowager Empress Ci-Xi (Tz'u-Hsi), rose against the 'foreign devils'. The rising was put down bloodily. War in South Africa between Britain and the Boer republics gave the century an early example of swift guerrilla tactics (by the Boers) keeping at bay a far larger but slower army, with famous sieges at Ladysmith and Mafeking set against an economic background of who controlled the diamond industry and the new goldfields spreading across the Rand. Kitchener defeated the Boers but in doing so herded many of their families into an early version of concentration camps where many died owing to food shortage and medical neglect. Some consciences in the Western world were awoken by Kitchener's actions as they were by the harsh regime imposed in the Congo through, in effect, the personal ownership of that country by Belgium's King Leopold. Gandhi's early agitation against racial policies towards Indians in South Africa also came to world notice.

In 1906 an earthquake tore San Francisco apart, though with much less loss of life than one of Europe's worst earthquakes at Messina in Sicily 2 years later. The huge task of rebuilding San Francisco began quickly. Such was the devastation, however, that many American insurers sought to limit the claims. Cuthbert Heath's renowned telegram to his agent in California—'pay all our policyholders in full irrespective of the terms of their policies'—was to set the seal on the leadership of Lloyds of London in the world insurance market for the next 50 years.

Transport and communication saw particular progress. The trans-Siberian railway was completed and the Hejaz railway in Arabia begun. The first electronically powered underground trains appeared in London and New York and Berlin built its first underground railway. The Zeppelin airship and the early aeroplanes of the Wright brothers and Blériot flew. The great motor companies of General Motors, Rolls-Royce and Ford—with its remarkable Model T (selling at $900)— were all founded (and, incidentally, disc brakes and car seat belts invented). Henry Ford showed his genius for cost-saving allied with an eye to marketing by saying that anyone could buy a Model T in 'any colour, so long as its black'. Building the Panama Canal began and the river Nile's Aswan Dam was completed but a Channel Tunnel Bill was aborted by Parliament in London.

While some of the ruling imperial families seemed trapped like flies in amber within 19th-century horizons, the discoveries of Einstein, Planck and Freud changed concepts of the physical universe and understanding the human mind. Marconi sent a radio signal across the Atlantic and Fessenden transmitted voice by radio for the first time. Villard discovered gamma rays, Bayliss and Starling hormones, and Landsteiner the blood groups A, B and O. John Fleming designed the diode valve that converted radio waves into electronic signals; this became important for radio and television. Heaviside traced the presence of an atmospheric layer that had consequences for the later development of radar. Nobel Prizes began to be awarded in 1901 and were won by the Curies and Becquerel for their work on radium and radioactivity, by Koch who identified the bacillus causing tuberculosis and by Pavlov for his work on the digestive system though he—and his dogs—also became famous for his discovery of conditioned reflexes responding to stimuli. It was the body's endurance as well as the mind's strength which carried Peary on to reach at last the North Pole.

The aeroplane

Wilbur Wright and Orville Wright, Dayton, Ohio
Filed 23 March 1903 and published as US 821393

Wilbur Wright was born in 1867 and Orville Wright in 1871. The brothers ran a printing shop together before switching to running a bicycle repair shop and later designing bicycles. The revenue helped to support them when they worked on aeronautical research. Wilbur became interested in flight when he read about Otto Lilienthal's fatal accident while experimenting with gliding in 1896. At the time much research into flight involved emulating the flapping of birds' wings. In 1899 Wilbur was watching buzzards in flight and realised that, besides using gliding, they twisted their wings to turn to one side. Flight control was vital besides the obvious need for propulsion. An aeroplane had to be able to bank, climb or descend and to steer left or right. Two or all three of these activities had to be done simultaneously.

The brothers resolved to sort out the problems of flight control before thinking of propellers and a light engine. They wrote to the Smithsonian Institution asking for material on aeronautical research and read all they could find. In 1899 they designed a biplane kite which had wings that could be mechanically twisted so that one wing had more lift and the other less lift. They then designed three biplane gliders during 1900–02, using a wind tunnel in Dayton to help their research.

The actual flights were tried out at a beach called Kitty Hawk, North Carolina. It was chosen after the Weather Bureau supplied them with a list of windy sites. The sand would protect the gliders from damage and the loneliness gave them privacy. The final version of the gliders had rear rudders for going to the left or right, forward elevators for going up and down, and the wings could be warped. Once they were happy with the gliders they designed a propeller and built their own 4 cylinder, 12 horsepower engine. The illustrated patent was filed 9 months before the first flight by Orville (chosen on the toss of a coin) on 17 December 1903. The beginning of the 12-second flight was photographed as was much of their previous research.

They then tried to sell the aeroplane to the American, French and British armies. Large sums of money were asked for but no demonstration was offered and they met with disbelief. It was not until 1908 that they began demonstration flights, having previously been scared of espionage, and the world realised that manned flight was possible. In a few years European aviation had surpassed their efforts. Wilbur died in 1912 and Orville in 1948. They both remained bachelors, flight being their only passion.

No. 821,393. PATENTED MAY 22, 1906.

O. & W. WRIGHT.

FLYING MACHINE.

APPLICATION FILED MAR. 23, 1903.

3 SHEETS—SHEET 1.

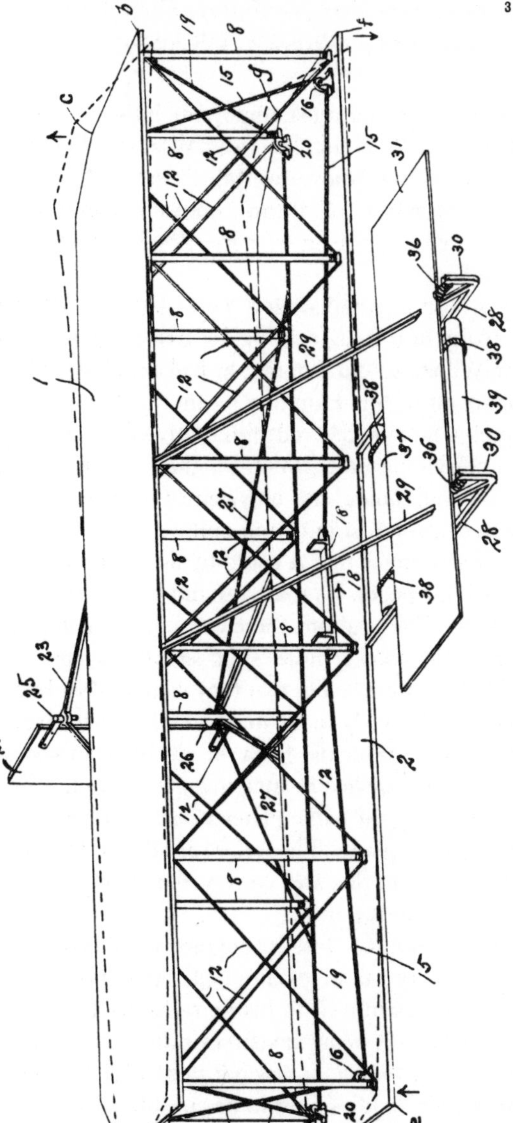

FIG. 1.

WITNESSES:

William F. Bauer.

Irvine Miller.

INVENTORS.

Orville Wright

Wilbur Wright

BY H. A. Toulmin.

ATTORNEY.

ANDREW B. GRAHAM CO., PHOTO-LITHOGRAPHERS, WASHINGTON, D. C.

Air conditioning

Controlling the temperature and humidity in air
Willis Carrier, Buffalo, New York for Buffalo Forge Works
Filed 16 September 1904 and published as US 808897

Air conditioning is perhaps the most important influence in making businesses function in hot climates such as the American West and South. The basic idea of using water to cool the air was known in Roman times. The Romans noticed that cooled vapour rose when water was thrown onto hot stones. In the 19th century, fans were sometimes used to drive air over ice. Besides temperature, humidity and dust had to be controlled.

Willis Carrier was born in Angola, New York state, in 1876. He worked for the Buffalo Forge Company in the engineering department. One day in 1902 a Brooklyn printer, Sackett-Williams, told him that he had a problem with colours blurring, since changes in temperature and humidity meant that the paper would expand and contract, so that each colour registered differently. That problem was sorted out with the first air conditioning unit, which weighed 30 tonnes, and the patent followed. The term 'air conditioning' was only coined in 1906, by Stuart Cramer, who added a dust filter to control dust in cotton mills.

The patent specifically states that it could be used for ventilating buildings and for other 'commercial' purposes. Its main use did in fact turn out to be on factory premises for many years. For example, there was a problem in a South Carolina cotton mill in 1906 where the 5000 spindles spun so fast that for several minutes after they ceased operating they were so hot that they would inflict burns if touched. Another use was in drying off newly made macaroni. In other cases as in textile mills it was important to remove impurities from the air. Figure 1 shows a fan B drawing the air into the trunk with a device H spraying water into the air. Next the air enters series of vertically arranged baffle plates where particles in the air are thrown by centrifugal force onto the plates. The water runs down the baffles through a sieve into a receptacle. Pipe-coils in this area convey a medium which can be adjusted if it is necessary to heat or cool the air.

Carrier's company decided to close its engineering department in 1914, apparently not having faith in the new invention. Carrier, his business partner-to-be Irvine Lyle, and four others moved to his new Carrier Engineering Corporation. Initially they concentrated on being inventors rather than manufacturers as they continued to improve the product. It was not until after World War II that prices came down enough for air conditioning to be considered a standard item in housing, so that now over 70% of American homes have air conditioning. In 1965 four massive Carrier units were installed to cool the Houston Astrodome. Willis Carrier died in 1950 in New York City with over 80 patents on air conditioning in his name.

W. H. CARRIER.
APPARATUS FOR TREATING AIR.
APPLICATION FILED SEPT. 16, 1904.

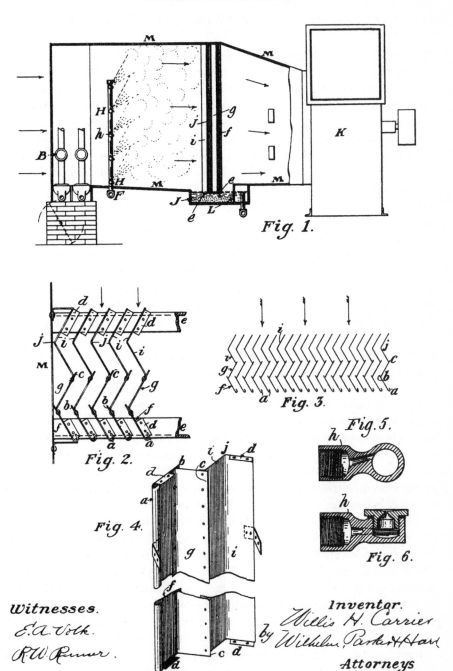

Fig. 1.

Fig. 2.

Fig. 3.

Fig. 4.

Fig. 5.

Fig. 6.

Witnesses.
E. A. Volk.
R. W. Renner.

Inventor.
Willis H. Carrier
by Wilhelm, Parker & Hart
Attorneys

The catalytic converter

Removing pollutants from internal combustion engine vehicle exhausts
Michel Frenkel, Paris, France
Filed 17 April 1909 and published as FR 402173 *and* GB 9364/1909

Many will be surprised that catalytic converters date back to 1909, but that is the year the patent was filed by a French chemist. There can be other types, such as ceramic beads, but the usual type is a honeycomb of ceramics or steel within a steel housing. The honeycomb is coated with a very thin layer of platinum, rhodium or palladium. The exhaust passes through and is enriched with air to create chemical reactions between the exhaust and the catalysts so that oxygen is added to the pollutants. Carbon monoxide and hydrocarbons are largely removed and the result is by-products of carbon dioxide and water vapour. A second chamber with a different catalyst is sometimes added in modern models to remove nitrogen oxides as well. Three grammes of the precious metals is all that is necessary.

Frenkel's invention covers the basic idea. He has a kaolin (china-clay) honeycomb with 30 grammes of platinum as the precious metal. He talks of 'deodorising' the exhaust with the aid of air being blown in with a fan. The idea behind that seems to be to ensure better combustion and making the exhaust less unpleasant, as if he were unaware of the key role that oxygen played. Figure 1 shows the device along its length, with Figure 2 a view from one end and Figure 3 a cross-section. The invention was not taken up at the time, perhaps because of the apparent need for as much as 30 grammes of platinum, or perhaps because there was not perceived to be a need for such an invention.

The first modern patent is perhaps US 3441381 by Engelhart Industries, filed in 1965. Nowadays metal is more often used than ceramics for the honeycomb as it has some advantages such as thinner walls in the honeycomb. The concept was given a huge boost in the United States with the passing of the Clean Air Act in 1970. Suddenly progressively cleaner exhaust emissions were necessary. Catalytic converters were first used in American cars in 1975. There are, however, some problems to their use. Lead destroys catalytic converters, so lead-free petrol was necessary before they could be used. Another problem is that converters need to be hot before they begin to work. This limits their effectiveness in cool climates and makes them useless on the many short journeys that drivers make, particularly as emissions are highest when the engines start up. Estimates are that a 5-kilometre journey is necessary before they begin to work. Also, converters become corroded and have to be replaced every few years. Finally, converters which work on nitrogen oxides give off nitrous oxides, a powerful greenhouse contributor. As the Clean Air Act is about smog and not greenhouse emissions there is nothing that the Act can do about it. Catalytic converters are now compulsory in most developed countries. The best solution to the problems of catalytic converters, of course, is to replace the internal combustion engine altogether.

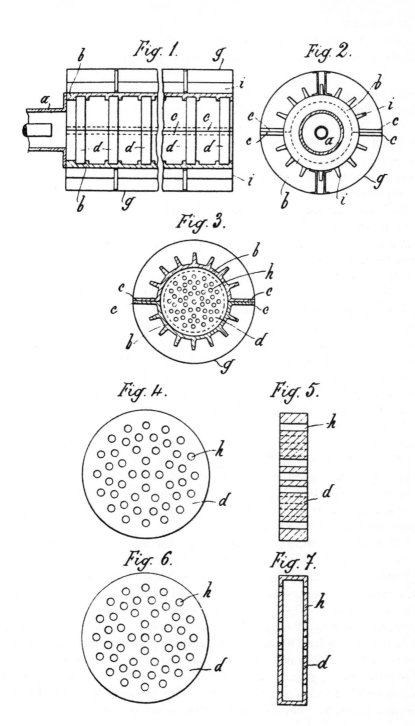

Fig. 1.

Fig. 2.

Fig. 3.

Fig. 4.

Fig. 5.

Fig. 6.

Fig. 7.

Disc brakes

Vehicle brakes
Frederick William Lanchester, Birmingham, Warwickshire, England
Filed 1 December 1902 and published as GB 26407/1902

This is the story of an 'unsung hero' of the automobile industry. When automobiles were first introduced the brakes used were brake shoes, similar to those used on carts. This was despite the fact that automobiles were heavier and faster than carts, and hence needed better brakes. In 1899 drum brakes came into production, which were manipulated by a hand lever. This involves a connection to a strip wound round the inside of the hub, which tended to lose efficiency with continual use owing to the heat.

Frederick Lanchester was a British pioneer of the aeronautical and automobile industry. He was Britain's most prolific inventor, with over 400 patents. Born in London in 1868, he was responsible for the building of the first British car in 1896. He became the chief engineer and general manager of the Lanchester Motor Co. He made costly cars which had little vibration, were smooth-running and which had the distinctive 'Lanchester look'. His GB 7909/1900, although about suspension, gives some idea of the appearance of such a car. Unlike the usual long bonnet, his cars looked bug-eyed as the engine was arranged length-ways between the front seats, with the petrol tank underneath the driver's seat. Lanchester pioneered the live axle and worm drive, thought out the idea of second gear and implemented a gear system close to automatic transmission, all before 1900. He admitted that he was better as an inventor than as a manager, and his ideas were always under-financed, so relatively little was actually achieved.

As an aside, perhaps, Lanchester lectured on the vortex theory of lift in aeronautics in 1894 to a Birmingham society and in 1897 offered a revised version to the Royal Society and the Physical Society, which both rejected it. It was only later realised that he had understood a vital idea before manned flight had been achieved, and his ideas were not printed until 1907. Both of these original texts were lost. He also wrote in the fields of radio, acoustics, relativity, music, poetry, operations research and military strategy.

Lanchester in his patent talks of gripping jaws like 'dentist's pliers'. Figure 1 shows a side view, Figure 2 a front view and Figure 3 a view from above. The disc (marked a) is riveted to the hub b. The jaws are c and d. When the lever p is pulled on by the driver, jaw c moves to grip disc a between it and jaw d. This causes normal rotation of the disc and hence the hub to which it is attached to stop. Modern disc brakes are based on the ideas in GB 688382 which involved hydraulic brakes with pistons, filed for in 1950 by Dunlop. Disc brakes were used on Jaguar's C and D types which won several races at Le Mans after which they were put on the company's production cars. Disc brakes are less prone to failure from overheating than drum brakes and at last became widely accepted over half a century after Lanchester's original design was patented. Lanchester himself had died in Birmingham in 1946.

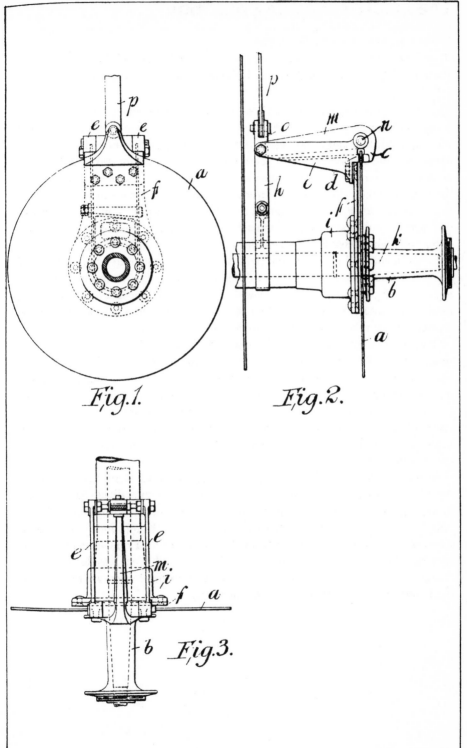

Fig.1.

Fig.2.

Fig.3.

[This Drawing is a reproduction of the Original on a reduced scale.]

Malby&Sons Photo-Litho

The electric washing machine

Electric washing machine for clothes
Alva Fisher for Hurley Machine Company, both Chicago, Illinois
Filed 27 May 1909 and published as US 966677 *and* GB 22114/1909

The idea of a washing machine had been around for a long time but these had been crude mechanical efforts which involved hand operating a stick within a box. This was not much fun, and a shortage of domestic labour did not help. In a way, the machine sounds eerily modern. The patent states that a 'perforated cylinder is rotatably mounted within a tub containing the wash water'. A series of blades lifted the clothes as the cylinder spun. After a number of revolutions in one direction, there were a number of revolutions in the opposite direction. This was to prevent the clothes 'wadding up into a compact mass' and was a good observation. The outer body which contains the cylinder is shown in the figure as 14 which is secured to an end wall 15. At bottom right is the electric motor 50. Drive belts from the motor leading to three wheels of different sizes operate the machine, including a clutch 53 at bottom left 'to throw the washing machine in and out of operation'. The long rod 60 was used as an emergency stop. Although the patent does not seem to be very clear it appears to be a 'top loader' model.

The machine was marketed as the Thor. Washing powder was invented just in time, with Persil® being the first of its kind, by Henkel of Germany in 1907. Fisher went on to patent other inventions in the same field, including two 'agitating mechanisms', gearings and a safety mechanism for clothes wringers. He also patented a gas burner. Improvements in the washing machine were slow to come until the market expanded, as many households initially lacked electricity. By 1929 84% of American households had electricity, but Britain was well behind in electrifying its houses.

In 1924 the first spin–drier was manufactured by Savage Arms Corporation of New York. Twin tubs first appeared in 1957. A fundamental divide has occurred in washing machines between top-loading machines, where a vertical fin sits in the middle of the drum and the clothes are loaded in from the top, and front-loaders, where a drum without any fins is loaded from the front. Front-loaders use just over half the amount of water that top-loaders use, and hence less detergent and energy. They are the main type used in Britain, while top-loaders are the norm in the United States, Australia and other countries.

A. J. FISHER.

DRIVE MECHANISM FOR WASHING MACHINES.

APPLICATION FILED MAY 27, 1909.

966,677.

Patented Aug. 9, 1910.

4 SHEETS—SHEET 1.

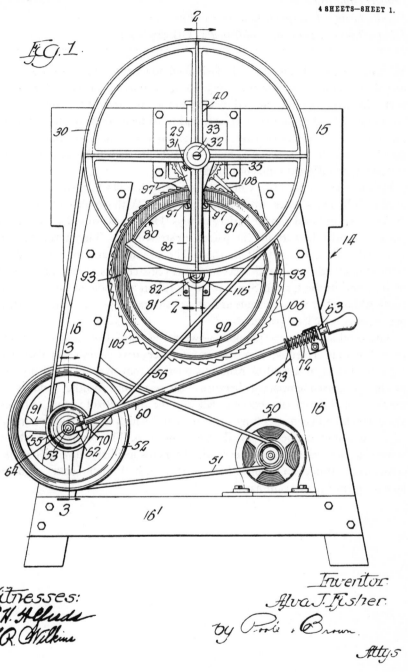

FIG. 1.

Meccano®

Construction toy
Frank Hornby, Liverpool, Lancashire, England
Filed 9 January 1901 and published as GB 587/1901

Meccano® is considered rather old-fashioned now but for decades it was very popular. Frank Hornby was born in 1863. He worked as a bookkeeper for a meat importer in Liverpool. He enjoyed making simple mechanical toys for his children, but found it frustrating that every toy had to be made from scratch with no simple components readily to hand. On a train journey one Christmas Eve he wondered if it would be possible to make versatile pieces which could be used again and again for different kinds of models.

He came up with a variety of metal strips with holes in them which could be used to fit brackets, axles, wheels, nuts and bolts. With simple tools they could be readily used to build a great variety of toys. Hornby borrowed £5 from his employer, David Elliott, to file the patent illustrated here. They formed a partnership and a factory was built in 1907. The original name of the product was 'Mechanics made easy', and Meccano® was registered as a trade mark in Britain in 1907 and in the United States in 1911. Elliott was bought out by Hornby in 1908. Although it was a struggle at first, sales began to pick up. Many further improvements were later patented.

The product was very popular between the wars. Coloured pieces first appeared in 1926. Hornby, who became a Conservative Member of Parliament in 1931, died in 1936, a wealthy man. His relatives took over the company. In 1942 production ceased temporarily because of the war effort. As other products caught the fancy of children sales began to fall, and the Binns Road factory in Liverpool, a site which had been operating since 1914 and which was famous among generations of young engineers, closed forever in 1979. The company went into receivership the same year. However, the French subsidiary, now an independent company, continues to produce Meccano® from its French factory, and another factory operates in Argentina.

A.D. 1901. Jan. 9. N.º 587.
HORNBY'S Complete Specification.

1 SHEET

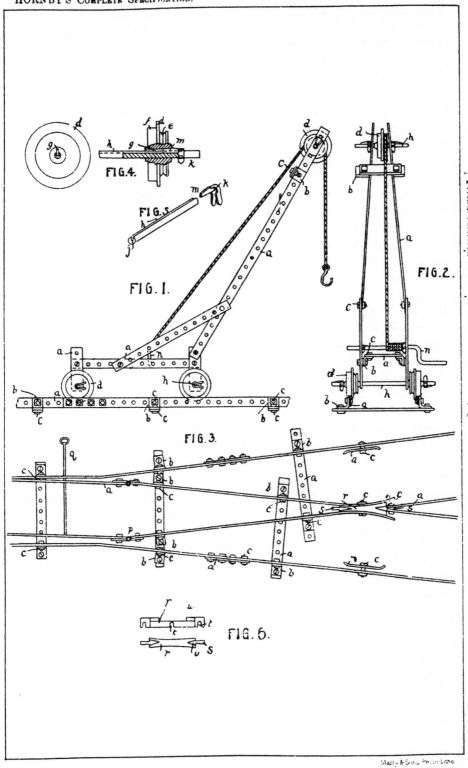

FIG.4.

FIG.5.

FIG.1.

FIG.2.

FIG.3.

FIG.6.

The safety razor

Razor which reduces the risk of cutting oneself
King Camp Gillette, Brookline, Massachusetts, for Federal Trust Company
Filed 3 December 1901 and published as US 775134-5 *and* GB 28763/1902

Gillette was born in 1855 in Wisconsin. His family soon moved to Chicago, where they lost everything in the great fire of 1871. The young Gillette tried to earn a living as a travelling salesman. He also came up with some minor inventions as a sideline. He began selling a new idea, cork-lined bottle caps. One day he met its (successful) inventor, William Painter. In their conversation, Painter suggested that Gillette invent something that was used, thrown away and bought again—perfect for a salesman.

The idea of a razor with a cheap disposable blade came to him 'in a flash' one morning while shaving. It was a risky business shaving yourself, with imminent danger of cuts, and the blades themselves had to be repeatedly sharpened, or a barber had to be visited. Gillette went out and bought pieces of brass, steel ribbon used for clock springs, a small hand vice and some files. He built an initial, primitive safety razor. For six years he worked on the idea. He needed to make a cheap blade from sheet steel that would harden and temper suitably to take a keen edge. He knew nothing of steel (and indeed had little engineering background) yet was confident that a viable product could be made when the experts said that it was impossible.

He managed to find some financial backers. The syndicate included an inventor named William Nickerson, who suggested making the razor handle heavy enough to facilitate accurate adjustment between the edge of the blade and the protecting guard. By 1902 endless experimenting resulted in determining the proper size, shape and thickness, a process for making the proper steel, a T-shaped handle so that it could be turned around for use on both sides, and equipment for making and sharpening the steel. The firm was by now in debt, but sales soon took off. In 1903 168 blades were sold but in 1904 over 12 million were sold. Gillette's own face was on the wrapper of each one.

Gillette soon became a millionaire and retired from active management in 1913, although he stayed on as President until 1931. He moved to California to grow fruit and to spend more time on his other passion, establishing a new economic order. Since 1894 he had been writing about abolishing wasteful competition and allowing engineers to run the world. There would be huge communal dining halls to eliminate the waste of each household cooking its own food, and Niagara Falls would power all industry. In 1910 he offered ex-President Theodore Roosevelt $1 million to head his World Corporation in the then Arizona Territory. King Camp Gillette died in 1932 in Los Angeles.

K. C. GILLETTE.
RAZOR.
APPLICATION FILED DEC. 3, 1901.

NO MODEL.

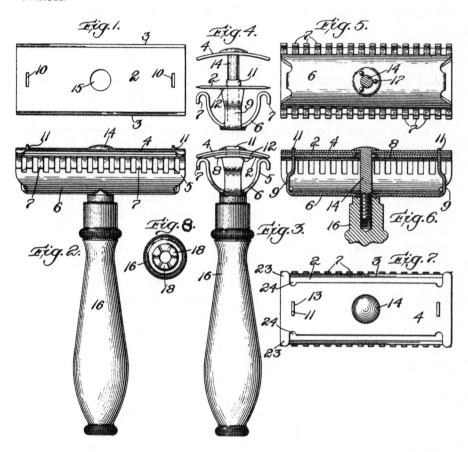

Tarmac® road surfacing

Mixture of slag or macadam with tar for surfacing
Edgar Purnell Hooley, Nottingham, England
Filed 3 April 1902 and published as **GB 7796/1902** *and* **US 765975**

It was John Loudon McAdam, Scottish engineer and General Surveyor of Roads in England from 1827, who introduced improved roads made of crushed stone, called 'macadamised' roads. These roads were a great improvement on dirt roads, which easily became impassable when it rained heavily. They were however still difficult to negotiate because of the stones, and punctures could easily occur on the tyres of the newly introduced automobiles. In 1901 Edgar Hooley, the county surveyor of Nottinghamshire, noticed that a stretch of road at Denby, Derbyshire had no ruts on it. He wondered why it had such a good surface, and asked if anything was special about this road.

He was told that a barrel of tar had accidentally fallen off a dray. In order to cover the tar, slag from the blast furnace at a nearby iron works had been added on top. He realised that here he could take a by-product of an industrial process and make something useful from it. The illustration from the patent shows the process of heating the tar, adding slag or macadam to the mix and breaking the stones within the mixture. Any leftover tar dropped down at the bottom. In 1903 Hooley formed the TarMacadam (Purnell Hooley's Patent) Syndicate Ltd, and registered Tarmac® as a trade mark in May 1903.

Hooley may have come up with a good idea, but he was not a good businessman. Sir Alfred Hickman was the owner of a large steelworks at Wolverhampton. He bought the rights to the patent and moved the location of the factory from Nottingham to Wolverhampton, and mercifully shortened the company name to Tarmac Ltd in 1905.

The company was successful, and when during World War I road-building ceased in Britain, it shipped vast quantities of material to France to build roads through the battlefields. After World War I the company began to buy up slag tips to meet the demand for so-called 'metalled' roads. During World War II the material was again much in demand, and 5 million tonnes were used to make roads or airport runways. The company continues to thrive today, and the trade mark is widely although erroneously used in Britain to indicate a road surface.

E. P. HOOLEY.
APPARATUS FOR THE PREPARATION OF TAR MACADAM.
APPLICATION FILED NOV. 3, 1902.

NO MODEL.

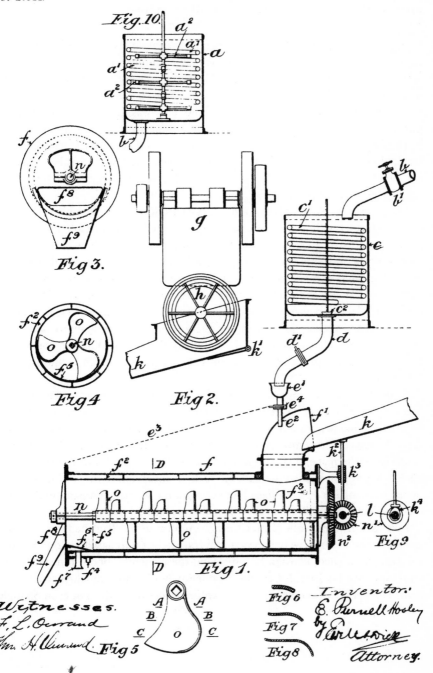

Fig. 10.

Fig 3.

Fig 4.

Fig 2.

Fig 1.

Fig 5.

Fig 6

Fig 7

Fig 8

Fig 9.

Inventor:
E. Purnell Hooley
by
Attorney.

The vacuum cleaner

Suction cleaning device
Hubert Cecil Booth, London, England
Filed 30 August 1901 and published as **GB** 17433/1901

Booth was an engineer who visited a demonstration at the Empire Music Hall in London. An American was showing his new cleaning machine which blew dust away from itself into a collecting box. The result was not very good. Booth asked him why he didn't change it to a suction machine so that the dirt could be more effectively drawn out and then trapped. The reply was that such a device was impossible. A few days later Booth was having dinner with friends in a restaurant. He had been mulling over the cleaning problem and suddenly placed his handkerchief over the antimacassar cover of the armchair and sucked as hard as possible. He nearly choked himself—but the handkerchief now had a patch of dirt on it, showing that the principle worked.

Booth then developed his machine. Electricity was rarely available in houses so it was designed with a motor and a pump mounted on a 'portable frame or carriage' in the street, while a flexible hose went into the room to be cleaned. Workmen in white uniforms operated the machine. The pump was so big that it had to be taken around on a horse-drawn cart. It was also very noisy, so horses were easily startled. When the carpets under the thrones at Westminster Abbey were found to be filthy before Edward VII's coronation in 1902, the vacuum cleaner carried out an excellent job of cleaning them. The King was so impressed that he ordered vacuum cleaners for both Buckingham Palace and Windsor Castle. With this royal seal of approval, society organised tea parties to watch vacuum cleaning being done.

If Booth were so important, why do people frequently say that they will 'hoover' the carpet? James Murray Spangler was an asthmatic janitor working in a Canton, Ohio department store. He hated rising dust when he used a brush or carpet sweeper, and combined a small motor, a rotating brush, a pillowcase and a brush handle to make the first portable, if primitive-looking, vacuum cleaner with his US 889823, published in 1908. He took his invention to his cousin, who happened to be married to William Hoover, who ran a small company making saddles—a declining trade when cars were fast replacing horses.

By this time American homes were beginning to be electrified, so a growing number of households were potential buyers. Hoover purchased the rights and his company flourished, constantly improving the principle. Booth himself tried to rival him with his British Vacuum Company, but without much success, so the name Hoover is associated with the invention rather than Booth or even Spangler.

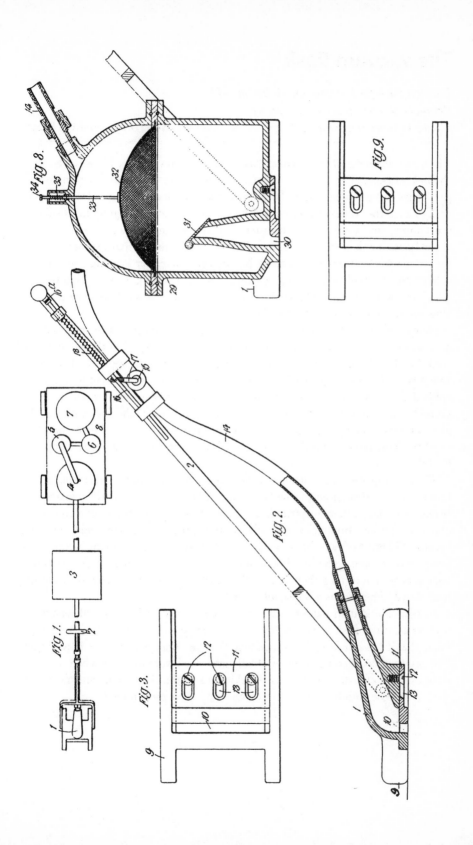

Fig.1.

Fig.2.

Fig.3.

Fig.8.

Fig.9.

The vacuum flask

Vacuum flask for keeping liquids hot or cold
Reinhold Burger of Berlin, Germany
Filed 1 October 1903 and published as DE 170057, GB 4421/1904 *and* US 872795

The actual inventor of the vacuum flask was Sir James Dewar. He was a Scottish chemist working at the University of Cambridge. In 1893, while working on the properties of very cold materials, he invented the idea of using a double-walled flask to keep the contents very cold. The fact that a vacuum is good for insulation was known, but by creating such a vacuum and inserting it as a layer round a glass vessel (glass being a poor conductor of heat), he was able to make something useful for his work. It would of course be effective for either keeping the contents cold or hot. It involved creating a vacuum and then sealing it by melting the top. The inner wall was coated on both sides with mercury. It was known to scientists who worked with it as the Dewar flask.

Dewar employed two Germans to blow the glass to make a flask, and used it in his teaching work. He did apply for a patent, GB 439/1893, but this merely covered the idea of removing air to make a vacuum rather than applying it to a suitable vessel. In 1903 one of the same German glassblowers, Reinhold Burger, applied for a patent. In 1904 he and the other glassblower formed Thermos GmbH. The name was suggested by a competition to name the flask where a Munich resident came up with the word from the Greek *therme*, hot. The name Thermos® was registered as a trade mark. Later Burger improved the idea by applying for DE 183666.

Three companies were set up in each country when rights were sold to the United States, Britain and Canada in 1907. The trade mark is still registered in Britain and many other countries but was lost in the United States in a 1962 court case with Aladdin Industries, a smaller manufacturer in the same field. This was because 'Thermos®' was being used as if it were a noun by many people, and hence had become 'generic'. The company's case was not helped by occasional poor supervision of how they treated the word: in a 1910 catalogue, for example, they stated 'Thermos is a household word'.

In 1910 the British patent was lost in a court case with Isola Ltd which was making a similar flask, and which counter-claimed, when charged with infringement, that it was not new. The judge held that the concept was in the Dewar flask. The product has of course been gradually improved from the days when the flasks were used on expeditions such as Shackleton's to the South Pole and Peary's to the North Pole, and by the Wright brothers on their early flights.

Fig. 2.

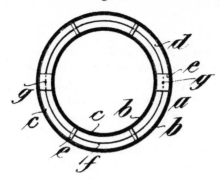

Fig. 1.

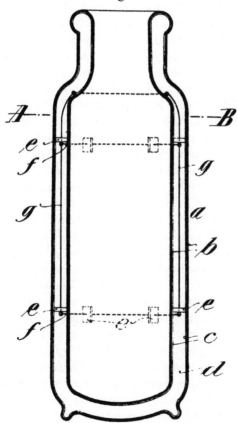

Zu der Patentschrift

№ 170057.

PHOTOGR. DRUCK DER REICHSDRUCKEREI.

1910–1919

THE tragedy of World War I (the Great War) overshadowed everything. Archduke Franz Ferdinand's murder unleashed the jealous ambitions and military competition of the great powers and brought to eruption nationalist tensions in the Balkans. The war redrew Europe's map, killed a generation, virtually bankrupted Europe's economies and ended the reign of many royal houses. Though the League of Nations was established by the peace conference of 1919 the harsh terms imposed by the Treaty of Versailles on the defeated nations meant, as Foch and Lloyd George feared, that it would prove not 'a peace treaty . . . [but] an armistice for twenty years'. The war created the circumstances for the Russian Revolution with its immense impact on the world. The USA, which had suffered less loss of life and economic resources but whose output in many industrial categories was now without equal, began after 1918 to represent a considerable influence in politics and economics across the world following its decisive entry into World War I. The scale of the Great War's battles made sure their names would endure: Verdun, the Somme, Gallipoli, Tannenberg, Ypres, Passchendaele, Gorlice-Tarnow and many more. This war introduced aerial bombing, attacks by submarine (or U-boat) and saw the 'Big Bertha' gun able to shell Paris from a distance of 120 kilometres. The employment of so many women in jobs vital for the war effort, adding to their campaigns for more equal rights, paved the way for recognition of greater equality for women and especially their right to vote: passed by laws in, for example, Britain in 1918 and the USA in 1920.

Efficient railways and an effective system of troop conscription, possessed for example by Britain and Germany but not by Russia, pointed the way towards the needs of later 20th-century warfare. But many military and psychological lessons, especially for the blood-stained stalemate of the Western Front, could have been learned from the American Civil War of the 1860s had Europe been concerned to study them. Masses of men hurled repeatedly against artillery, machine-guns, trenches and barbed wire defined the image of the Great War for later generations with unimaginative and distant generals deciding the lives of heroic, long-suffering soldiers: 'lions led by donkeys'. There were some exceptions. Winston Churchill and other military historians were to consider that the ablest general of the Great War, because of his intellectual grasp of strategy and sense of manoeuvre necessary for huge armies, was Germany's Max Hoffmann to whom (among others) the 'lions and donkeys' phrase is attributed. (For similar reasons and qualities some came to view Erich von Manstein as his counterpart in World War II. Both led armies in bitter and fluid fighting in the vast spaces of Russia.)

At Jutland on 31 May 1916 the British Grand Fleet and the German High Seas Fleet joined battle in the only major sea encounter involving modern fleets in European waters. The outcome was inconclusive. However the long myth of the Grand Fleet's invincibility was swept away at Jutland and, while German gunnery and some naval engineering proved superior, the High Seas Fleet never ventured from its home ports again to seek to break the British sea blockade. Thereafter German naval warfare, as in World War II, would largely involve U-boats. Casualties among British seamen at Jutland numbered 6,000: on 1 July 1916, on only the first day of the battle of the Somme, the British army suffered 60,000 casualties (20,000 killed).

Some 10 million people lost their lives in the Great War, more if all those 'missing' could be counted (compared with perhaps 55 million, including an immeasurably greater number of ordinary citizens, in World War II—20 million of them people of the Soviet Union). From 'official' rough estimates, including some allowance for those missing-in-action probably killed (Kipling's 'known unto God'), Russia lost almost 2.5 million soldiers and military personnel, Germany more than 2 million, Austria-Hungary 1.75 million, France 1.5 million, Britain and its Empire 1 million, Italy more than 0.75 million, Romania and Turkey nearly 0.4 million each, and the USA almost 150,000. More than twice as many were wounded. But, as ever, a war such as that of 1914–18 produced its own innovations: chlorine and mustard gas (chillingly evoked in Wilfred Owen's poem *Dulce et decorum est*) and Fokker's synchronised machine gun to fire through an aeroplane's propeller blades among them. And the tragedy advanced medicine: Mott's outline of the war neurosis of shell shock and Gillies' initial work in plastic surgery (to be carried further by McIndoe in World War II).

The origins of World War I involved imperial rivalry and nationalist aspiration, not—as some were to come to see World War II—a crusade against evil and the slaughter of helpless innocents. A nationalist's bullet fired by Gavrilo Princip of the Serbian 'Black Hand' prompted the Great War, Berchtold the Austrian Foreign Minister began it, Kaiser Wilhelm II could have prevented it. 'Man proposes, but God disposes': after the man-made carnage of 1914–18 twice as many as those killed in the Great War died in the 'Spanish 'flu' epidemic—or pandemic—of 1918–19. Most of these died of pneumonia to which the most devastating influenza virus of the century had made people vulnerable: a far more virulent 'flu strain than the later visitations of 1957–58 and 1967. The Great War seems to have robbed the decade of time to invent more for peaceful purposes. But the years did achieve the development of the model of the atom and the general theory of relativity (Rutherford, Bohr and Einstein), the discovery of vitamins by Funk and McCollum's isolation of vitamin A, Ford's conveyor-belt assembly line for industrial production and Claude's invention of neon lighting. Alcock and Brown's non-stop Atlantic air flight provided a reminder of the triumph of transport's new technology after the loss of the 'unsinkable' ocean liner *Titanic* in 1912 had previously undermined confidence in it. If the decade closed with a peace treaty in Paris which would inflame further animosities in Europe it had begun with an event in Paris which would lead to long-term cooperation among nations—a first air traffic conference.

Formica®

Easy to maintain floor and tops laminate
Daniel O'Conor, Jr, Pittsburgh, Pennsylvania for Westinghouse Electric and
Manufacturing Company
Filed 1 February 1913 and published as US 1284432

Laminates are materials which consist of layers of materials, sometimes different ones. The first patent for a laminate was by Leo Baekeland, a Belgian working in America who had invented what became known as Bakelite®. He worked in the later stages with two engineers from the Westinghouse Corporation. These were Herbert Faber and Daniel O'Conor. Both were just out of college when they met in 1907. They had a mutual dream which they discussed at weekends: to invent a material that would change people's lives. The company's first effort at a laminate was impregnating heavy canvas with Bakelite®. Although the patent was filed through Westinghouse they thought that the company was not committed to the invention, and so they left and formed the Formica Corporation later on in 1913.

The material is typically 1.5 mm thick and consists of several layers of unrefined 'kraft' paper impregnated with phenolic acid which is compressed between heated sheets of polished steel. This is then bonded to wood, chipboard or plywood.

Formica® got its name from 'for mica' because it was thought of as a substitute for mica, the mineral that was widely used for electrical insulation, but which was expensive. It was used in cars and in radios and was made initially on premises rented cheaply in Cincinnati. It was not until the mid-1920s that the company thought that the material might be useful in furniture. At this time the material was available only in black or brown but they introduced a method of lithographing the top surface with designs, such as wood grain or marble effects, and it quickly became popular with restaurants and bars. It was easy to clean and maintain for a start; it was also cigarette-proof, which was much appreciated. Shiny black Formica® was popular with designers working in the modern style and this variant was used in New York's Radio City Music Hall in combination with chrome and aluminium.

In 1937 melamine began to be used to impregnate the top layer to increase the hardness and water resistance. This also made it easier to produce lighter colours. In 1957 the company hired Raymond Loewy to design many avant-garde designs on a white background that were very popular with many for worksurfaces in the kitchen. It was about this time that Formica® moved out of commercial premises into the home and became a familiar trade name. Although much liked, it is probably to be an advertiser's dream that some housewives, delighted with the ease of use of the product, sang 'Formica, Formica, oh how we like ya'.

UNITED STATES PATENT OFFICE.

DANIEL J. O'CONOR, JR., OF PITTSBURGH, PENNSYLVANIA, ASSIGNOR TO WESTING-HOUSE ELECTRIC AND MANUFACTURING COMPANY, A CORPORATION OF PENNSYLVANIA.

PROCESS OF MAKING COMPOSITE MATERIAL.

1,284,432.   Specification of Letters Patent.   **Patented Nov. 12, 1918.**

No Drawing.   Application filed February 1, 1913.   Serial No. 745,616.

To all whom it may concern:

Be it known that I, DANIEL J. O'CONOR. Jr., a citizen of the United States, and a resident of Pittsburgh, in the county of Allegheny and State of Pennsylvania, have invented a new and useful Improvement in Processes of Making Composite Material, of which the following is a specification.

My invention has particular reference to methods of manufacturing composite materials, such as cardboard.

One object of my invention is to provide an insulating material which is light, strong, and insoluble and has a high dielectric strength.

Another object is to provide a simple and efficient process of producing an insulating material of the above-indicated character in large quantities as a commercial product.

Heretofore, insulation material such as cardboard, composed of layers of paper glued together, has proved more or less unsatisfactory because of various defects, such as absorption of moisture from the atmosphere, inability to resist heat and chemical action, and lack of physical strength. Insulating material to be used in connection with switchboards and wireless telegraph and other high voltage installations, must be free from these defects, and, in addition, must possess high dielectric strength. My invention provides a process of manufacturing an insulating material possessing these qualities in a high degree.

In carrying out this process, any suitable fabric, such as paper, muslin, or other cloth, and fibrous or porous material of any kind may be used. If paper is used, the fabric preferably baeren paper, kraft paper or the so called micafolium paper which has a coating of mica flakes, of any desired thickness, is first coated on one side with an adhesive liquid insulating material, suitably that known as bakelite and consisting of a condensation product of phenols and formaldehyde. The coating operation is performed by passing the paper between two rollers, the bottom one of which dips into the liquid material which is contained in a tank. The thickness of the coating retained by the paper is regulated by varying the distance between the two rollers and by altering the viscosity of the liquid. The paper is then dried by passing it over a series of rollers in a steam-heated oven. The prepared paper is cut into sheets of any desired size but, for convenience, preferably 18"x36" or 36"x36", as desired. A plate is built up to the required thickness by placing the sheets together with the untreated side of each sheet next to the treated side of the adjacent sheet, the number of sheets required for any desired thickness of finished material having been previously determined. The upper sheet is preferably placed with its treated side down, in order that both the top and the bottom of the finished plate will present untreated faces.

The built-up plate is then placed between thin sheet steel plates on which has been rubbed a small amount of machine oil. Any desired number of the steel plates carrying the sheets of paper are placed between the platens of a hydraulic press which have been previously heated, preferably by steam. The press is closed and pressure applied, which may be as high as 800 pounds per square inch, or approximately, 535 tons on an area 36"x36". Satisfactory results have also been obtained by using lower pressures. Heat is applied, preferably by steam, while the material is in the press. The pressure is kept constant during the period of heating and the subsequent period of cooling. These periods are varied according to the thickness of the plate approximately in accordance with the following table:

Thickness of plate.	Time under steam.	Time of cooling.
Up to $\frac{1}{16}''$	15 min.	10 min.
$\frac{1}{16}''$ to $\frac{5}{8}''$	30 "	15 "
$\frac{5}{8}''$ to $\frac{7}{8}''$	45 "	20 "
$\frac{7}{8}''$ to $1\frac{1}{4}''$	1.00 hr.	25 "

The effect of heating and pressing the plate is to firmly cement together the sheets of paper and to further impregnate the paper with the bakelite. The plate is transformed into a hard and compact mass.

After cooling, the plates of insulation are removed from the press and then clamped between steel plates to prevent warping during baking, which is the last step in the process and which is employed for the purpose of removing any moisture remaining in the insulation and for transforming the bakelite completely into its infusible and insoluble condition. The plates are then placed in ovens, air pressure of approximately 140 pounds per square inch is applied, and the temperature is regulated be-

Neon lighting

Using rare gases to provide coloured illumination
Georges Claude, Seine department, France
Filed 7 March 1910 and published as **FR 424190**

Neon, a rare element found in the atmosphere, was only discovered in 1898. Named for 'neos', Greek for new, it is inert. This means that it does not react with other elements if mixed.

Georges Claude was a scientist working with gases who had formed his own company, Société l'Air Liquide, to exploit his many patents. He carried out an experiment where he conducted electricity though neon. He found that a bright red glow resulted. The patent involves taking a tube containing neon or other rare gases with electrodes on the outside. Additional gas could be added at long intervals through a valve. 'Neon' is really a misnomer as other gases can be added, or used, to vary the colours. Argon, another inert gas, will give a pale blue colour, for example. Claude was aware of this and the patent covers both using neon and a mixture of gases, or a tiny amount of mercury. He mentions using helium for yellow or white. Claude developed a process for obtaining neon from the atmosphere by liquefying air. He also invented non-reactive elements large enough to handle ion bombardment without heating or spluttering. These were all necessary for neon lighting to work.

The tubes themselves are simply formed by bending a stick of glass. The bombardment with high-voltage electricity via the electrodes at each end causes the gases to light up. The tubes are relatively cool unlike most lights which give off a lot of heat. The first commercial use was at a motor show in Paris in December 1910. With its bright colours it was realised that it was an ideal medium for advertising and this occurred at a barber shop on the Boulevard Montmartre. It was not until 1923 that neon lighting penetrated to the United States when Earle Anthony, a Los Angeles Packard car dealer, bought two back from a trip to Paris. The cost was similar to that of a small house. The bright colours immediately stopped traffic.

Neon tubes are also visible in poor conditions, have a long life of 20 years or more and are generally maintenance free. They are not good for reading, though. The flourescent tubes used in offices are a variant as they contain argon and krypton, and are better for reading. They were introduced in 1935.

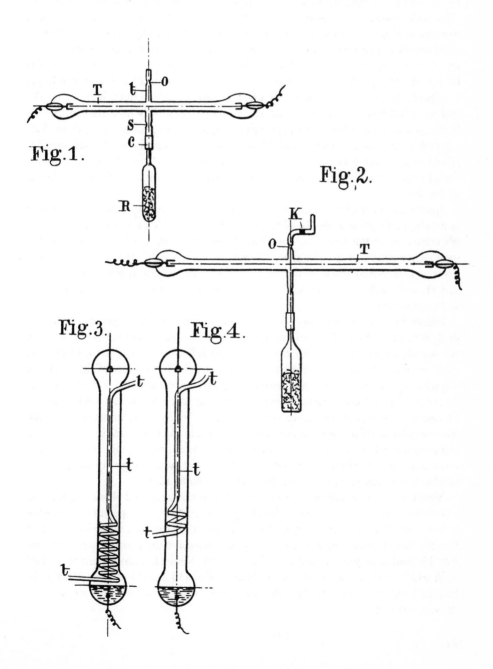

Fig.1.

Fig.2.

Fig.3.   Fig.4.

The Nissen hut

Portable building
Peter Norman Nissen, British Expeditionary Force, France
Filed 26 June 1916 and published as **GB 105468**

Peter Nissen was born in the United States in 1871, son of a Norwegian immigrant who made 'stamp-mills', or devices for crushing ore. In 1891 the family emigrated to Canada where Nissen studied mining at Queen's University, Kingston. He travelled widely as an engineer, but mainly in England, and patented several inventions, mostly for stamp-mills. When World War I broke out he joined the army although he was already 43. While serving on the Western Front as a Captain in the Royal Engineers, he noticed that there was a lack of simple, easy to build housing for the troops. What houses there were near the trenches had mostly been shelled. Nissen began to sketch ideas for his hut. He claimed later that the rounded roof of the skating rink at his old university inspired him with his design for the Nissen hut. He talked to his senior officers, and a general said that he was 'too valuable an inventor for a Field Company R.E.'. Three prototype huts were built to work out the details of the design.

Nissen was allowed to patent the invention in his own name, so the British patent names him a Captain in the Royal Engineers in France. He cautiously made his application care of the Institution of Mining and Metallurgy in London in case the mail didn't get through. The idea was not so much a building that could be dismantled and moved about, as one that could be shipped out in pieces, and erected quickly and easily. The semi-circular structure means that no heavy load-bearing walls are needed.

Nissen huts are built by first laying wooden beams along the ground to form a grid. Several semi-circular steel ribs are bolted to the beams, and horizontal wooden ribs are attached in turn to these ribs. The wooden floor is screwed to the beams and an inner lining of light corrugated iron is fixed behind the joins of the steel and wooden ribs. A second covering, with its distinctive corrugations, is then placed as an outer skin, and lastly vertical wooden boarding is erected at each end, with one end having a door, windows and a ventilation grill. Among other improvements that were suggested to him and adopted was the double lining to provide insulation. The standard hut is 8.2 metres long, 4.9 metres wide and 2.4 metres high. One hundred thousand huts were built by the end of the war, the parts being shipped out from England. It normally took six men four hours to build a hut.

Nissen patented a number of improvements to the structural elements, and also a steel tent, GB 129777. He was offered £500 for war use of his patent and rejected it, hurt, but later accepted £10,000 tax free and other royalties when it was found that the British government was shipping out his huts to the United States after the war. He died in 1930 at his home outside London. Nissen huts continued to be used in World War II, an American variant being called a Quonset hut. The main problem with the huts was the fact that they were cold in winter, and stoves were definitely needed.

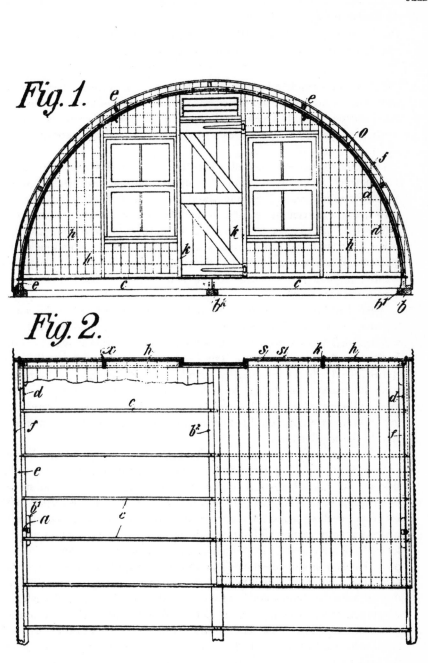

Fig. 1.

Fig. 2.

The outboard motor

Motor for use on a small craft
Ole Evinrude for Evinrude Motor Company, both Milwaukee, Wisconsin
Filed 16 September 1910 and published as US 1001260

Ole Evinrude's family emigrated from Norway to Wisconsin when he was five, in 1882. One of his earliest memories was spending most of the voyage exploring the ship's engine room. As a teenager, he spent much time tinkering about and reading mechanics' magazines. He set up a small company making automobiles at a time when they were a distinct rarity.

In 1906 Evinrude got engaged to his office manager, Bess Cary. One hot day they were having a picnic on an island 8 kilometres from the shore when Bess expressed a wish for an ice cream. Evinrude rowed all the way to shore to buy one. Although both strong and keen, he realised on the way back that it was not just automobiles which could use an engine, nor was the wet mush he offered up much appreciated either. He went to work and designed an 1125 watt engine with a weight of 28 kilogrammes. When he tried it out the following summer his wife said that it looked like a coffee grinder. Just the same it sold well, helped by Bess' slogan 'Don't row! Throw the oars away!'

There had been experiments with outboard motors before but this was the first one that really worked. It looks surprisingly like modern models. Made of iron, it was designed for easy dismounting and in the configuration shown, 13 is the clamp for attachment to vertical sterns. This can be moved to the right to allow for flaring sterns. The flywheel 8 is turned by a crank 9 to start the engine, 10 being the petrol tank. Evinrude later applied for US 1011930, a canoe powered by an engine mounted near the bow, which was not successful. His first engines were put on rowing boats and it was not until 1914 that boats were deliberately designed to be powered by outboard motors.

Evinrude formed a partnership with a tugboat owner called Chris Meyer. After years of working hard in the business, Evinrude in 1914 sold his rights to Meyer so that he could take a long break. A condition of the sale was that he did not work actively in the field for 5 years. At the end of 5 years, Evinrude returned to Milwaukee from travelling round the country and offered Meyer his new aluminium motor, which was twice as powerful but considerably lighter. Meyer said no, and Evinrude formed a new company, Elto, which competed with his old company. A third company, Johnson Motors, meanwhile took the lead in the field. In 1929 Evinrude took over his old company which became the Outboard Marine Corporation. Life was difficult with the Depression, but Evinrude helped out his staff with cash gifts while developing improvements such as electric starters and much more powerful engines. The company also diversified into lawnmowers. Bess died in 1933 and Evinrude, saddened by her loss, died the following year. The company continues to lead the field, though oddly the well-known trademark Evinrude® was only registered in 1949.

1,001,260.

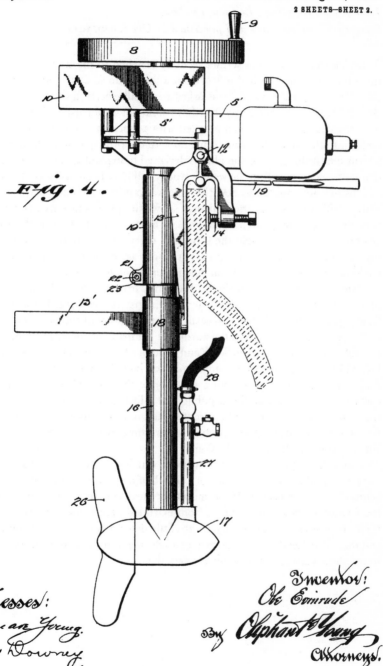

Fig. 4.

Witnesses:

Caravan Young.

May Downey.

Inventor:

Ole Evinrude

By Oliphant & Young

Attorneys.

The Rawlplug®

Plug for securing screws, etc. to walls
John Joseph Rawlings, London, England
Filed 14 November 1911 and published as **GB 22680/1911**

A mechanism for keeping screws securely fastened to the wall may seem a simple thing but someone still had to work out a way to do so easily. According to legend John Rawlings, a contractor, was asked to install electrical fittings to a wall in the British Museum in such a way that they would not be obtrusive. Traditionally a hole was chiselled, which was then plugged with a block of wood. A screw was then inserted into the wood. Rawlings came up with the principle of the Rawlplug®, where 'expansion means grip' as the plug is designed to expand as the screw is inserted so that it fits securely within the wall. It would still have been necessary to make a hole, and history does not tell us how this hole was put in, which nowadays is easily done with a drill. Figure 2 in the drawings shows the plug before use. It consisted of a tube weakened along longitudinal lines by being drawn from reels through a die. A coating of glue or gum held the little balls together. Figure 6 shows such a plug in use, where the plug has collapsed outwardly as the screw is inserted. The plugs could be made of a variety of suitable materials such as jute.

A single machine was installed at the family firm, Rawlings Brothers, to make the plugs, and the trade mark Rawlplug® was registered in 1912. It was fortunate for Rawlings that his trade mark so easily reminded users of the place where his plugs could be used, and shortly afterwards the company changed to its present name, Rawlplug Ltd. Soon sales took off and a factory had to be taken over. Demand was helped by a big advertising campaign which included taking an entire page in the *Daily Mail*: common today, but a sensation at the time. Demonstrations were also carried out to show a doubtful public that a little plug could be a stronger fixture than a large wooden plug. The patent should have run its full term of 14 years but was allowed an extra 4 years' protection, presumably because the company was unable to exploit it during World War I.

A later development was the Rawlbolt®, which was the subject of a patent filed in 1934, GB 444623. This consisted of a sheath of metal which pressed outwards as the central bolt was tightened. It was meant for use in concrete or masonry. In the 1960s the Rawlplug® began to be made from extruded plastic and is more sophisticated, typically having a slit end so that it can easily deform outwards, four slits at the base, and three little teeth to provide grip on each side near the end. Handymen continue to use these convenient little plugs every day.

A.D 1911. Oct. 14. N.º 22,680.
RAWLINGS' COMPLETE SPECIFICATION.

(1 SHEET)

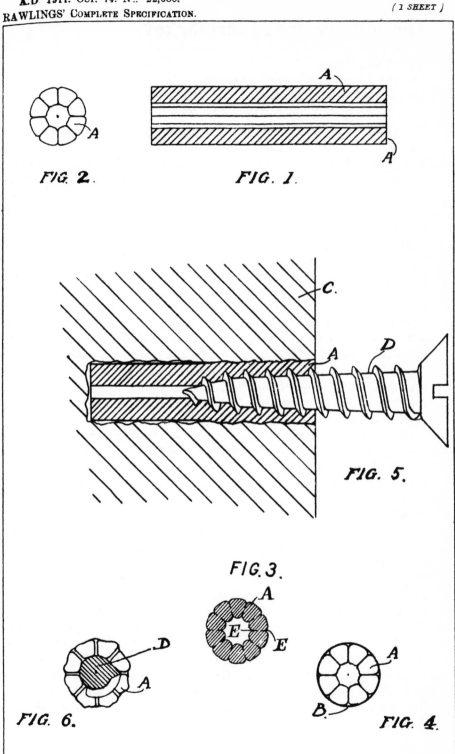

FIG. 2.

FIG. 1.

FIG. 5.

FIG. 3.

FIG. 6.

FIG. 4.

The self-service supermarket

Convenience store
Clarence Saunders, Memphis, Tennessee
Filed 21 October 1916 and published as US 1242872

It may sound strange but the idea of a supermarket was indeed patented, and by the man who thought of it. Clarence Saunders was a flamboyant and innovative man who thought about the waste of manpower and space in the conventional store of the time where customers had to ask staff behind a counter to hand over whatever was required from shelves behind them. His patents do not show an interest in helping customers to be able to browse or save their own time, but rather in making the operation of a store more economical for the owners.

Saunders opened the first store operating on his new principles at 79 Jefferson Street, Memphis, on 6 September 1916. The patent was filed a few weeks later. The splendid drawing opposite shows the front of the store. The customer enters the food display area on the left, through the gate below 'Aisle no. 1' and proceeds through four aisles (two of which are visible behind a screen) on a prescribed 'circuitous' path, to emerge to the right of 'Exit' and, going past a final display, doubles back to pass in front of the cash register and leave through a second gate. The patent states that 'The customer is required to review the entire assortment of goods carried in stock' before leaving after 'relieving the store of a large proportion of the usual incidental expenses, or overhead charges, required to operate it'. He states that three or four times the normal volume of sales can be dealt with by the same number of staff.

He named the new chain Piggly Wiggly® and franchised the format as high-volume, low-profit margin stores. The chain is still operating, mostly in the American South, with 600 stores. A legend about the unusual name is that he once saw from a train several piglets trying to get under a fence. A better explanation may be his reply when asked why the chain had that name, 'So people will ask that very question'.

Over the next few years Saunders patented several improvements to the concept of a self-service store, as well as such ideas as price tagging (his were the first stores to price every item), a lighting system for self-service stores and tape for adding machines. He was so successful that he was able to begin building a huge house in Memphis. Unfortunately, he went bankrupt after disputes with the New York Stock Exchange over a series of stock transactions, and never lived in the house. It is now the Pink House Museum, and contains a replica of his first supermarket. Saunders recovered sufficiently to open a chain trading under his own name, but this closed in the Depression. He began to work on an even more economical store, which would be fully automatic, but there were mechanical failures, though patents on this idea were published as US 2661682 and US 2820591, which came out after his death in 1953. He obviously would have loved the idea of selling over the Internet.

C. SAUNDERS.
SELF SERVING STORE.
APPLICATION FILED OCT. 21, 1916.

1,242,872.

Patented Oct. 9, 1917.
3 SHEETS—SHEET 1.

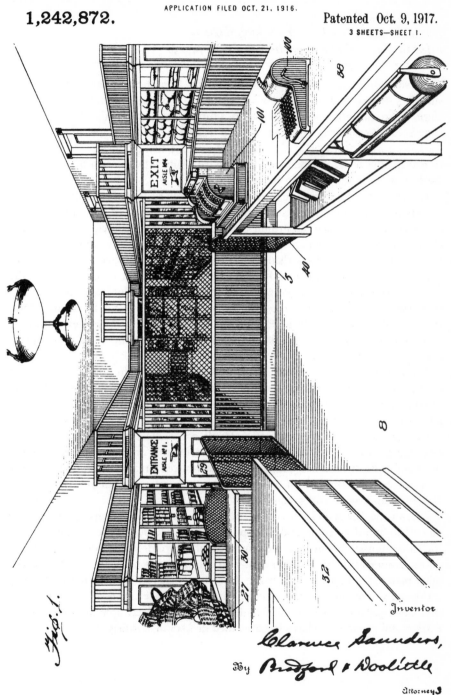

Fig. 1.

Inventor

Clarence Saunders,
By Bradford & Doolittle
Attorneys

Stainless steel

Rust-proofing of steel
Harry Brearley, Sheffield, Yorkshire, England
Filed 29 March 1915 and published as US 1197256

The evolution of stainless steel is confused and complicated: this can only be a summary, based on Harry Brearley's work. He was born in 1871 in Sheffield, Yorkshire, where his father was employed as a furnaceman for Firth's, the steel works concern. He began work at the age of 12 as a cellar lad and was later apprenticed to the laboratories. Studying at night school, he rose so that in 1907 he was appointed to manage the research laboratory that was run jointly by the two main Sheffield steel firms, Firth's and Brown Bayley.

In 1912 he was investigating the corrosion through rusting of rifles. Research had already been carried out into such properties. Alloys consist of mixtures of different metals where each contributes a useful quality. Stainless steel includes chromium in its alloy, but at least 12% of the steel must be chromium for the stainless qualities to show. Less than 1% of carbon is also essential. Léon Guillet of France published a detailed study of such alloys in 1904 but he failed to note the corrosion resistance quality. Portevin in 1909 also failed to note it, while an article by Giesen in 1909 contained the exact proportions used by Brearley in his patent. Philipp Monnartz and Wilhelm Borchers of Germany discovered and explained the anti-corrosion quality. They obtained DE 246035 for an alloy of 10% chromium and 2 to 5% molybdenum.

Brearley made a cast in the electric furnace with 12.8% chromium. After a heat treatment the resulting metal resisted corrosion. The government was not interested in using this new metal. Brearley suggested to his company that it be used to make cutlery. He asked a local cutler's to make knives for him (rusting steel knives when washing up were a big problem). He called it 'rustless steel' but one of the cutlery managers dubbed it 'stainless steel' after seeing that dropping vinegar on it did not stain it. Firth's did not want a patent but resisted Brearley's wanting to patent it which explains the lack of a British patent. Brearley left for the rival Brown Bayley's company as works manager.

Meanwhile, Elwood Haynes of Indiana was researching in the same area. His wife had asked him if he could make rustproof cutlery. He independently came across the chromium-steel alloy, and filed for a patent before Brearley, but it was rejected on the grounds that 'these chromium-iron alloys are not new'. In 1919 he did obtain a patent, US 1299404, for a 'wrought-metal article'. Another variety of the alloy was being worked on by Maurer and Strauss at the large German firm of Krupps; they produced an alloy in 1912. By adding nickel they were able to produce some extra useful properties. Harry Brearley died in 1948 in Torquay, Devon.

UNITED STATES PATENT OFFICE.

HARRY BREARLEY, OF SHEFFIELD, ENGLAND.

CUTLERY.

1,197,256. Specification of Letters Patent. **Patented Sept. 5, 1916.**

No Drawing. Continuation of application Serial No. 17,856, filed March 29, 1915. This application filed March 6, 1916. Serial No. 82,301.

To all whom it may concern:

Be it known that I, HARRY BREARLEY, residing at Sheffield, Yorkshire, England, have invented a certain new and useful Im-
5 provement in Cutlery, of which the following is a full, clear, and exact description.

My invention relates to new and useful improvements in cutlery or other hardened and polished articles of manufacture where
10 non-staining properties are desired and has for its object to provide a tempered steel cutlery blade or other hardened article having a polished surface and composed of an alloy which is practically untarnishable
15 when hardened or hardened and tempered. This alloy is malleable and can be forged, rolled, hardened, tempered and polished under ordinary commercial conditions.

The invention results from the discovery
20 that the addition of certain percentages of chromium and carbon to iron will produce a steel capable of taking a polish and having the characteristics above referred to. I have discovered that the addition to iron
25 of an amount of chromium anywhere between nine per cent. (9%) and sixteen per cent. (16%), and also an amount of carbon not greater than seven tenths per cent. (.7%) will result in a product which, when
30 made into knife blades, has the said characteristics.

I have further found from experiments that steels containing less than eight per cent. (8%) of chromium are relatively tar-
35 nishable whatever the amount of carbon that they contain up to the limit at which they cease to be malleable and capable of being hardened and tempered. I have also found that when the amount of carbon exceeds
40 seven tenths per cent. (.7%) the polished steel is tarnishable whatever the amount of chromium it may contain and that this condition corresponds with the appearance in the steel of free carbids, which are distin-
45 guishable microscopically on polished and etched specimens.

A typical composition for the untarnishable steel blades embodying my invention would be as follows: carbon .30 per cent.;
50 manganese .30 per cent.; chromium 13.0 per

cent.; iron 86.4 per cent. In producing such steel I preferably use an electric arc melting furnace. It can be readily made in such furnace. It forges easily into sheets or strips such as are required for knife 55 blades and can be hardened and tempered by ordinary commercial processes.

Knife blades embodying my invention are made from the steel above referred to being formed, hardened and polished by grinding 60 or buffing in the ordinary manner, the product being a polished cutlery blade similar in appearance to other polished blades but possessing the remakable quality of being practically untarnishable when subjected to the 65 ordinary uses to which knife blades are subjected, because made from the alloy above described. My blades are tempered so as to be sufficiently resilient for ordinary requirements. 70

Small amounts, up to say one or two per cent. of nickel, copper, cobalt, tungsten, molybdenum and vanadium appear to be without influence on the untarnishable property of the steel. 75

In practice it is best not to attempt to obtain an alloy containing above .4% of carbon, but rather to try to obtain an alloy containing an amount of carbon less than .4% thus leaving a wider margin for varia- 80 tions from the alloy sought to be produced since the desired result is attained when considerably less carbon is present.

This application is a continuation of my application Serial No. 17,856, filed March 85 29th 1915.

As is evident to those skilled in the art, my invention permits of various modifications without departing from the spirit thereof or the scope of the appended claims. 90

What I claim is,

1. A hardened and polished article of manufacture composed of a ferrous alloy containing between nine per cent. (9%) and sixteen per cent. (16%) of chromium and 95 carbon in quantity less than seven tenths per cent. (.7%).

2. A hardened, tempered and polished cutlery blade composed of a ferrous alloy containing between nine per cent. (9%) and 100

Synchronised machine gun fire from aircraft

Fighter aircraft with a machine gun designed to fire safely through propellers
Franz Schneider, Berlin, Germany
Filed 15 July 1913 and published as DE 276396 *and* GB 16726/1913

This invention had a tremendous impact on aerial combat in World War I, and could have won the war for Germany. Neither they nor the British were aware that the idea had been patented in both their countries, and was sitting on library shelves waiting to be consulted.

Fighter aircraft were still in their infancy in 1914, and the only way of bringing down an enemy fighter was to take a pot at him with a rifle or a pistol. Some aircraft carried an observer behind the pilot, who would try to aim and fire a heavy machine gun held in his arms, which was awkward when his lap was full of bombs. Just before the war, French manufacturer Raymond Saulnier had been working on an interrupter gear that would allow a machine gun to be fired through the propeller arc. Unable to work out how to do it, he attached steel deflection plates on the propeller so that it would be unaffected (although the pilot might be). The military lost interest in his idea once the war started. After a few months of war, the pilots were crying out for fixed machine guns facing forward so that they could both aim and shoot in the direction they were flying. French flier Roland Garros, who had been a stunt pilot before the war, came to Saulnier and had steel deflector plates attached to his propeller blades, and a fixed machine gun mounted in front of the cockpit. Garros was able to shoot down five German planes in a fortnight. On 19 April 1915, though, he was brought down by a rifle shot over Courtrai. His attempts to set fire to his plane were unsuccessful, and the modified airplane was quickly in the workshop of Anthony Fokker, the great Dutch aeronautical engineer, who was working for the Germans.

Tests were carried out on the aircraft, but the heavier German steel-jacketed bullets destroyed the propellers. In two days Fokker worked out a way of using gears to link the crankshaft of the engine to the firing of the machine gun, so that the bullets would avoid hitting the propeller. His DE 299770, which was only applied for on the 28 June 1916, may be for an improvement on the idea. Franz Schneider promptly charged him with infringing his patent, which admittedly shows a rifle rather than a machine gun. There were only two blades to propellers, so only two pauses were needed within each revolution. The only earlier work on the subject, FR 463601, filed in January 1913, was not feasible. It features a very modern looking fighter plane shooting down a dirigible, with the gun barrel coming out of the centre of the propeller itself rather than behind it, or from the wings. It was by Louis Blériot, the first man to fly across the English Channel.

Fokker Eindekkers were armed with synchronised Spandau machine guns and roamed the skies virtually unopposed for a period known as the 'Fokker Scourge'. In April 1916 the Allies in their turn captured a Fokker aircraft and discovered the secret, and they soon deployed their version of the new system to the front, although initially 'accidents were common'. The Germans continued to improve the mechanism, and numerous German and British patents were published after the war on the subject.

Fig. 1

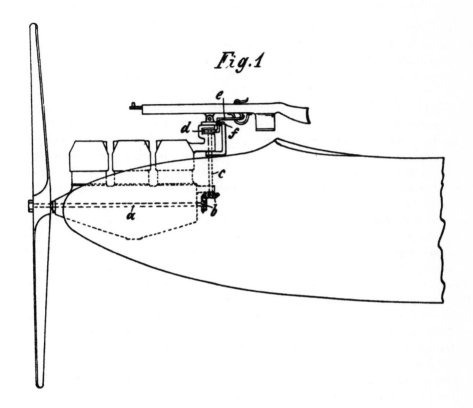

Fig. 2

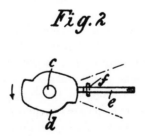

The tank

Armoured gunnery vehicle with endless tracks
Gunter Burstyn, Korneuburg, Austria
Filed 28 February 1912 and published as DE 252815

The origins of the tank are shrouded in mystery. Traditionally it is attributed to the British but Burstyn's invention illustrated here seems to precede the British work. Admittedly, it does not feature an endless track mechanism. The popular story about using tanks for warfare is that Sir Ernest Swinton suggested the idea of an armoured vehicle with moveable tracks to Lord Kitchener at Christmas 1914 as a way of getting round the impasse of trench warfare. Machine guns were preventing advances by infantry, and trenches, shellholes and mud were blocking progress by vehicles. A committee was formed in February 1915 by the Admiralty, the Landships Committee. They asked for proposals for an armoured vehicle equipped with heavy guns which could cross trenches and other obstacles. One (unsuccessful) proposal by a naval pilot was for three 12.2 metre wheels being used to drive a huge machine, a veritable leviathan.

Soon William Tritton, the managing director of a Lincoln agricultural machinery company (William Foster's), got involved. He began filing patents from May 1916 onwards and these were published as eight patents from GB 125105. They all involved the idea of endless or 'caterpillar tracks', which to us may now seem obvious in a tank, but did not mention their use in an armoured fighting vehicle. The first model in September 1915, Tritton No.1, shed its tracks too readily but a modified machine worked better and was used as a working model to plan changes. It was nicknamed 'Willie'. He was joined by an army engineer, Walter Wilson, in further work. By December 1915 the 'Mother' model was ready. This was the familiar beast seen in old photographs with huge tracks on each side. They could cross 3-metre wide trenches. The army ordered 100 of them and had them called water tanks for security reasons, and the name stuck.

In March 1916 a special arm of the army, later to become the Tank Corps, was formed. Tanks were first used in a minor action on 15 September 1916, but their first major use was at the Battle of Cambrai on 20 November 1917, when 474 went into action. They caused consternation and many prisoners were taken, but in fact many tanks either broke down or were crippled by artillery fire. The tank was not really that successful during the war but it was useful as propaganda. Tritton and Wilson were awarded compensation after World War I by the British government for the use of their inventions. Some tanks were made for the French by Renault while the Americans used Ford copies of the Renault tanks. The Germans did produce the A7V tank but did not believe in the concept. Perhaps the course of the war would have been changed if they had adopted Burstyn's idea at an early stage.

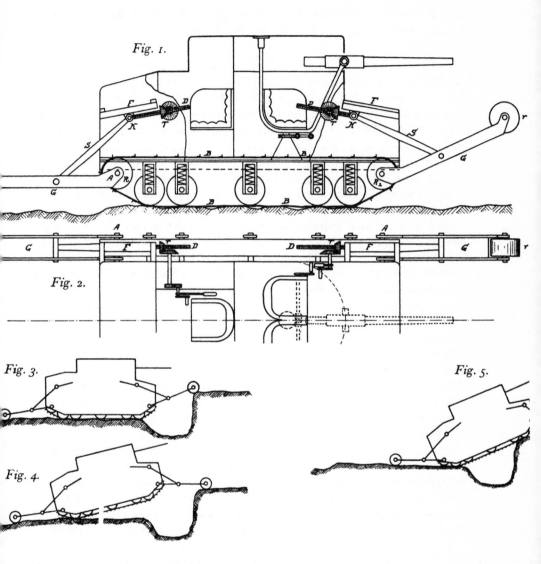

Fig. 1.

Fig. 2.

Fig. 3.

Fig. 4.

Fig. 5.

The zip fastener

Gideon Sundback, Meadville, Pennsylvania, for the Hookless Fastener Company
Filed 27 August 1914 and published as US 1219881 *and* GB 12261/1915

Traditionally either buttons or some sort of hook and eye arrangement were used for closures in clothes, bags etc. where zip fasteners would frequently be used today. Whitcomb Judson's US 504037–38, published in 1893, were the first for the concept of a chain of clasps that could be closed or opened by moving a guide along the seam. Designed for shoes, it was bulky and impossible to mass-produce easily. It also had a tendency to 'pop open' at inconvenient moments. Backed by 'Colonel' Lewis Walker, a lawyer, Judson continued to work on the idea. His US 1060378 was thought foolproof on the popping open problem, only for the Company Secretary to need a safety pin when trying them out one evening. Work went on until 1908.

Then Gideon Sundback came on the scene. He was born in Sweden in 1880 and trained as an electrical engineer, emigrating to the USA in 1905. He came to the company for an interview and stayed on, marrying the daughter of Peter Aronson, one of the mechanics there. He was to receive all the foreign rights to any patents.

So far work had been on the idea of hooks and eyes. Sundback followed this line (meanwhile sadly his wife died in childbirth) but had no success and then decided to try something quite different. Jaws would be on one side, beaded edges on the other. A slider would either press them together so that they snapped shut or pull them apart. It was not quite right (the beaded strips wore out easily) but it was felt that these were on the right lines. Finally a second version had metal teeth on one strip which nested left or right into similar teeth on the other strip, again with the slider. As the two strips were identical and interchangeable the concept could be mass produced, and although it was improved to make it cheaper and easier to make it was the basic idea of the zip fastener. Sundback himself devised machines for stamping out the parts and fastening them to the tape.

The company had difficulty stirring up interest from manufacturers who might buy quantities for their products. World War I helped as they were used in money belts, flying suits and life preservers. Then the Goodrich Company, which made rubber products, asked for a huge number and launched a boot, the Mystik Boot. Goodrich's company president is reported to have said when salesmen reported that the name did not attract interest, 'What we need is an action word. . . something that will dramatize the way the thing zips'. Then he realised that he had it. They registered a trademark for 'Zip fastener boot' in 1925 but the word became so popular that it was used as a noun by most people, and so eventually lost its protection. It was not until 1935 that the idea of using a zip fastener in clothes became popular.

1,219,881.

Patented Mar. 20, 1917.

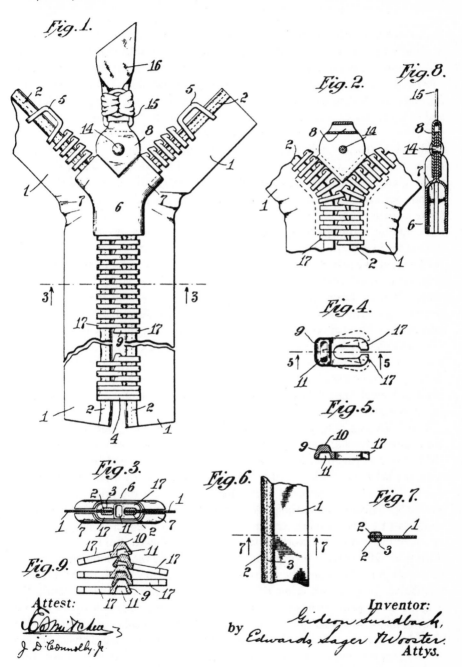

Inventor:
Gideon Sundback,
by Edwards, Sager & Wooster.
Attys.

Attest:

1920–1929

HAND in hand with the images of the jazz age and the roaring '20s, conjured up in some Western societies, walked the dark legacy of World War I bringing with it economic breakdown and the collapse of currencies. Scott Fitzgerald's novel *The Great Gatsby* evoked some of the atmosphere and picture of post-war gaiety for those who had money and to whom stock-market booms brought recklessness and extravagance '. . . careless people . . . [who] smashed up things and creatures and then retreated back into their money or their vast carelessness . . .'. The wealth of this society, especially in the USA and in the affluent circles of Europe, contrasted starkly with those whom World War I had dispossessed or disadvantaged. Unemployment facing those returning from war did not make Britain 'a fit country for heroes to live in' and their dissatisfactions led eventually to a general strike in 1926.

Rebuilding a Europe broken by war was slow and new ideology found fertile ground for its growth within the economic chaos: Fascism in Italy and Nazism in the German Weimar Republic where inflation ran at 2,500% each month at the height of the 1923 crisis. Following Lenin's death political struggles on the future shape of Communism ensued in the Soviet Union. From this Stalin emerged as the most powerful figure setting in train (and to be enforced and increased brutally in the 1930s) collectivisation programmes especially for agriculture designed to turn the Soviet Union into a modern state. In China, Chiang Kai-Shek became leader of a modernising nationalist government beginning conflict with Mao Zedong's communist movement which would simmer and erupt for over 20 years. Mustafa Kemal Atatürk began to build the modern state of Turkey by defeating the invading Greek army in 1922.

The USA began to pursue an isolationist international policy through the 1920s. But increasing sales of products and new innovations within business in the USA, owing in part to mass-production techniques, led to wide prosperity and a stock market boom with record levels of bank lending. By 1922 increased production and sales of improved Ford Model Ts had cut the price of each car to around $300 (from $900 in 1908). Since Ford from 1914 had guaranteed his workers at least $5 a day, buying cars and the new goods came increasingly within the means of many ordinary working families. Money flowed into the stock market often from safer areas of investment. At the same time the prohibition of the sale of alcohol in the USA and the flourishing illegal industry which resulted laid the foundations of organised crime there, and elsewhere, for decades to come. The long market in rising shares on the USA stock exchange ended in 1929 when the value of major stocks fell by over two-thirds and many bankruptcies and the loss of lifelong savings followed. This caused further international damage because much European money had been invested in booming markets in the USA. Both capitalist and communist systems were grappling with economic distress.

Nor were the ravages of economics the only contagion with which nations had to deal. The great 'flu epidemic and its consequences at the close of World War I, which affected half the world's population, coincided with another disease from a different virus but for which the influenza virus paved the way by weakening people's resistance. Beginning in 1916–17 and enlarging in the 1920s an unprecedented pandemic of sleeping-sickness assailed the world, killing or maiming almost 5 million people before it ceased unaccountably in 1927. Some victims survived for many years in a coma or trance-like state. In a sign of the century's medical progress, but also of the journey of medical achievement still to be travelled, a very few of these sufferers of *encephalitis lethargica* who had survived to middle or old age were to be awoken—even if only briefly and fitfully—half a century after their illness through the use of the drug L-Dopa by Oliver Sacks.

The decade saw considerable progress in flight technology including an early form of helicopter (autogyro) and liquid fuelled rocket. An international flight (by Pan-Am) was launched, the first in-flight movie shown and notable individual achievements of flying recorded: Lindbergh's Atlantic solo flight, Byrd's flight over the South Pole, Earhart becoming the first woman pilot to fly across the Atlantic (also flying it solo in 1932). Britain's first national airline, Imperial Airways, was formed in 1924. Campbell and Seagrave both achieved car land speed records of over 320 kilometres/hour while, at the other end of the scale, the first motor scooter was manufactured. In 1928, in a notable milestone for rail transport, the 'Flying Scotsman' express train ran non-stop for more than 600 kilometres between London and Edinburgh. Television was demonstrated and both the British Broadcasting Corporation and the Radio Corporation of America came into being. The first UK-USA radio broadcast was made. Marie Stopes opened a pioneering birth control clinic in London, Goldberger isolated vitamin B and Banting the hormone insulin. Tytus's method of continuous hot-strip rolling for steel was invented and such diverse but influential products as the first insecticide, Kodak's 16 mm colour film, Kalmus's technicolour movie film and John Thompson's sub-machine gun were produced. In an important invention for future needs Geiger and Müller produced a much improved 'geiger' counter to detect radioactivity.

The autopilot

Apparatus for automatically keeping level flight and correct bearings
Frederick William Meredith for the Royal Aircraft Establishment,
Farnborough, Hampshire, England
Filed 18 July 1925 and published as GB 365186–7 *and* 365189–90

The history of the autopilot is riddled with claims and mystery. Elmer Sperry is thought to have invented the first working autopilot in about 1910 and to have used it in a Curtiss flying boat in about 1912 but this does not seem to have been patented. Another early claim is the Aveline stabiliser which was fitted to a Handley Page Transport in January 1921. Frederick Meredith led a team at the Royal Aircraft Establishment working on the problem of keeping airplanes steady without the intervention of a pilot. Originally the hope was that it would be the core of a system for pilotless bombers.

Gyroscopes were thought of as the key to any such system. These are spinning bodies set in a gimbal mount which respond readily to any change in course or elevation because of the use of three axes. Two gyroscopes are used, one to control the bearing or direction, the other to keep in level flight. If either change the gyroscope senses it, and a piston valve controls compressed air to a servomotor of which the piston is coupled with cables in the rudder, which responds to correct the error. In practice this model could only be used in calm weather.

Because of the possible air weapon use the patents were kept secret for several years. Then a press release was sent out and the *Illustrated London News* and the *Graphic* were invited to write a story, which came out in August 1930. The illustration in the article showed a luxurious airliner with two airmen calmly smoking and drinking while the plane flew itself. The idea was that the patents would be published and filings for foreign patents would be made immediately. In fact the patents were only published in December 1931 and it was not until July 1932 that patents were applied for across the world. Unfortunately records show that the foreign patents cost more to apply for than came in in royalties, as the world (or the airplanes) was not yet ready for widespread civilian air travel.

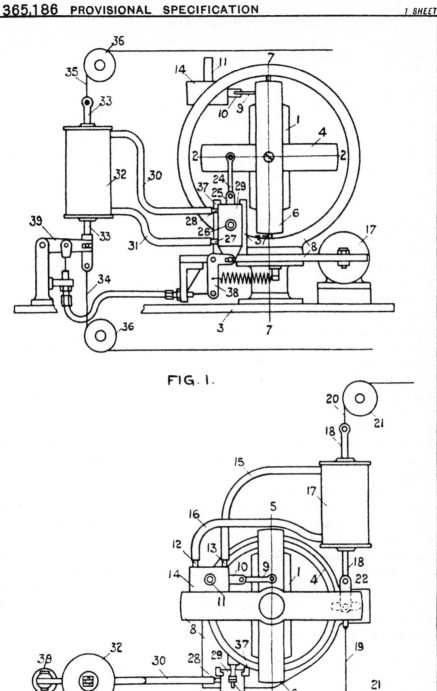

FIG. 1.

FIG. 2.

The bread-slicing machine

Machine for slicing and wrapping bread
Otto Frederick Rohwedder for Mac-Roh Sales and Manufacturing Company,
both Davenport, Iowa
Filed 26 November 1928 and published as US 1867377

Otto Rohwedder a jewellery store owner, began work on a machine for slicing bread in 1912. He asked bakers if they liked the idea, and they replied that sliced bread quickly went stale. He realised that a secure wrapping would be needed to keep moisture in, and thieving hands out. In 1915 he was told by doctors that he had one year to live. In 1917 he lost a prototype and all his tools and equipment in a fire. He also went bankrupt. It was not until 1922 that he actually secured financial backing for his ideas.

Figure 4 shows the front of the machine. The numerous parallel lines are endless cutting bands made of high grade spring steel. The apparatus shown in Figure 5 guides and drives them on the axis 5 to 5 at each end of Figure 4. The loaf moves between in the rectangle shown by 33 and adjacent bands move forward to slice in opposite directions to prevent crushing of the slices. A conveyor belt moves the loaf in and then takes away the slices between side belts to where they can 'cooperate' with a wrapping machine. To allow for different thicknesses of the slices, the screws 36 (at right) are turned which adjusts the arms 40 and 41 and hence the bands.

The new machine was first used at Battle Creek, Michigan and by 1933 80% of American bread was sold pre-sliced, an astonishingly quick rise. The new convenience food was greatly liked, particularly in combination with an electric toaster, which fortunately had already been invented by Frank Shailor with his US 950058 in 1909. Pre-sliced bread has made fewer inroads in the British market and even now, many loaves are sliced in front of the customer in the supermarket rather than at an industrial bakery.

Rohwedder was definitely a man obsessed with his subject. All of his many patents seem to be involved with bread. His US 1591357 is a display rack for bread while US 1724368 and US 1759592 involve using staples to hold sliced loaves together, a rather frightening thought. There are various devices for handling and feeding loaves, and US 1935996, US 2034250 and US 2061315 are further thoughts on slicing bread machines. The last of these was filed in 1935 and involves slots for receiving the cutting bands. Even US 1740038, a device for making wires, is almost certainly to do with making the cutting bands. The expression 'the best thing since sliced bread' apparently only dates from the 1950s as army slang and so, surprisingly, was not coined at the time of Rohwedder's invention.

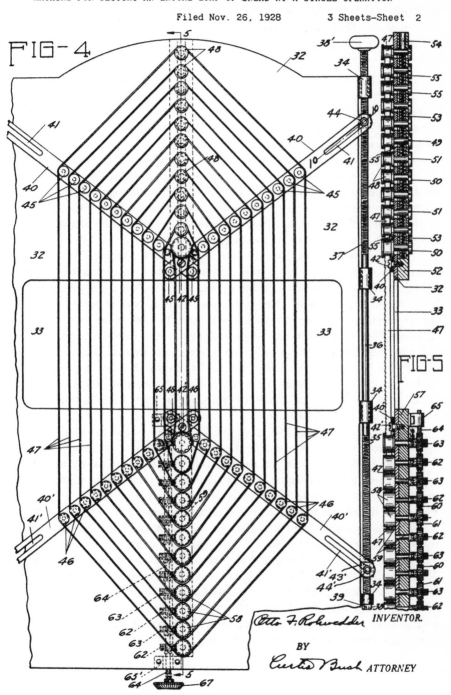

FIG-4

FIG-5

Otto F. Rohwedder INVENTOR.

BY

Curtis Bush ATTORNEY

Power steering

Enabling easy turning of vehicles by using hydraulics
Francis Davis, Waltham, Massachusetts
Filed 11 May 1927 and published as US 1790620

In the early days of automobiles and other motor vehicles some thought was given to assisting the driver in turning the steering wheel, but this did not become important until heavy lorries and buses began to be manufactured in the 1920s. It was tiring to make turns as so much effort was made in turning the wheel, especially when manoeuvring to park. Francis Davis was the chief engineer of the truck division of the Pierce Arrow Motor Car Company. He understood these difficulties and thought that, as this would be a constantly growing problem as the industry developed, it would be profitable to devise a solution. He read through relevant literature on the subject and decided that a hydraulic system was best. He wanted to keep any profits to himself rather than with the company so he left and hired a small workshop to work on the problem in conjunction with a tool-maker. He also obtained advice from a professor at the Massachusetts Institute of Technology.

Davis made his original application in April 1926 but refiled and five patents in all were published in his name during 1931 to 1933. He had already demonstrated his system in October 1926. The opening page of the illustrated patent is a discussion of problems which he proposed to overcome. Besides the problem of the 'exhausting effort' for the 'operator', he mentions the extra difficulties caused by the vogue for large balloon tyres which made steering even more difficult. He also discusses what he called 'reversibility', the need to ensure that shocks caused by bad road conditions would not jar the driver holding the steering wheel, resulting in 'wheel fight'. Complete 'irreversibility' would not be a good idea as then the operator would lose any feel for the road.

The basic idea was to use hydraulics to drive the direction of the wheels. An (oil) fluid-powered piston is connected to the wheels and engages with a nut which has a threaded engagement with the steering column. Hence by gently turning the steering wheel the piston does the hard work of turning the wheels. Figure 1 shows the general picture, while Figure 2 is a sideways view of the power unit 10. Figure 3 is an enlargement while Figure 4 is a view across 4–4 on the left of Figure 2. The valve was open while the mechanism was not in use. The other patents cover the servo-motor, auxiliary power steering gear and minor improvements to the mechanism, such as providing a drainpipe to take away excess fluid. After the demonstration, General Motors agreed to take out a licence on the technology. The Depression, however, caused them to abandon plans until 1941 when the United States entered World War II and plans were postponed again. After the war the great demand for cars made them think again and it was not until 1951 that Chrysler introduced the technology for passenger cars.

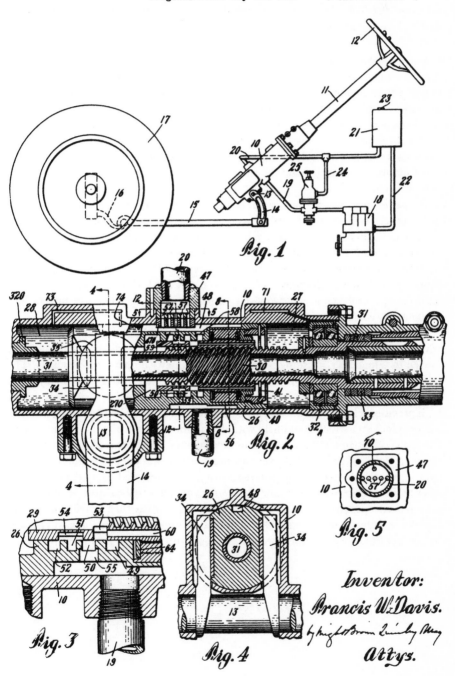

Fig. 1

Fig. 2

Fig. 3

Fig. 4

Fig. 5

Inventor:
Francis W. Davis.
by Wright Brown Quinby Otcey
Attys.

Rapid freezing of food

Quickly freezing food for preserving and later defrosting
Clarence Birdseye, Gloucester, Massachusetts for Frosted Food Company
Filed 18 June 1927 and published as US 1773081 *and* GB 292457

Clarence Birdseye was born in Brooklyn in 1886. He worked first as a naturalist and then as a fur trader in Labrador during 1912–17 and there he noticed that the Eskimos, when catching fish, would place them on a piece of ice. In winter, exposed to the icy winds, they froze almost immediately, but this was less effective when it was less cool, and the more quickly frozen fish was noticeably better-tasting. The fish could be eaten months later. He could watch the fishermen for hours. What happens is that fast freezing means that only small ice crystals are able to form. The cell walls are not damaged and the frozen food, once thawed, keeps its flavour, texture and colour.

Birdseye's first patent, filed in 1924, US 1511824, was titled 'Preparing piscatorial products' so he was still only thinking of fish. He commented in it that the presence of air, as in barrels of iced fish, hinders freezing and allows bacteria to flourish once temperatures rise again. It involves an insulated chamber with numerous cells or 'frames' which hold fish with the space between filled with brine acting as a refrigerant. This was a laborious way to try to freeze food. Birdseye was also responsible for US 1608832 which proposed something startlingly like fish fingers (but without any coating).

He set up a company but it went bankrupt. The public knew about cold stores, which as they froze the food less rapidly resulted in food that did not taste good. He moved to Gloucester, a fishing port, and set up a new company with backers. The illustrated patent shows his improved technique. The 'belt froster' has two icy metal belts with packages of food travelling between them. The food is flash frozen under high pressure. He still couldn't sell the product and decided he needed a partner in the food industry which could help with distribution. Eventually the Postum Company, a cereal manufacturer, agreed to be a partner. He later sold out his existing rights to them. This included using his name which was kept as a trade mark, but split in two. Birds Eye® was registered in 1931. Postum tried using demonstrators offering free samples in shops, and speakers giving talks to local groups, and slowly sales picked up. Diversification was made into other areas such as fruit juices, but vegetables were a problem until Donald Tressler realised that blanching was needed before freezing. Clarence Birdseye died in 1956 with over 350 patents to his name.

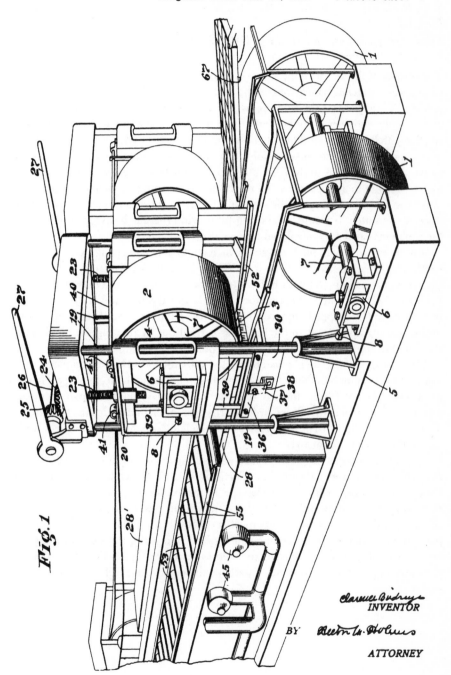

Fig. 1

Clarence Birdseye
INVENTOR

BY

ATTORNEY

The self-winding wrist watch

Watch which automatically winds itself without batteries
John Harwood, Lonon, Isle of Man and Harry Cutts, New Brighton, Cheshire,
England
Filed 7 July 1923 and published as **GB 218487**

The origins of the self-winding watch go back to the 18th century. These, however, were special watches which were easily damaged, difficult to repair, bulky and expensive. They were of the fob type which were kept in the pocket. By the 1920s most people wore wrist watches rather than the fob type as it was possible to make smaller watches which could easily be worn on the wrist. They were kept going by winding a crown on the outside. John Harwood was a watch repairer. He was unaware of previous work in the field, but wanted to develop a self-winding watch. His motive was not to help people avoid the trouble of winding a watch but rather to remove the crown, which gave dirt an opportunity to get into the movement and to damage the watch. It would also be helpful to wind the watch up regularly without the problems caused by misplaced human effort.

Within a year he had made a watch which could generate enough power to wind itself. The illustrations are of the front (Fig. 1) and back (Fig. 2) of the watch. A tiny swinging weight (A) with buffers (A1) was actuated by the movement of the wearer to swing through an arc of 300° to hit both sides. (C) and (C1) are stops. A friction plate (D) stopped any possibility of over winding. The absence of a crown meant that an adjustment to the time was carried out by rotating the bezel round the face. He applied for patents in several countries including Switzerland.

The task of making such a watch was too difficult for a British company, so he visited Swiss companies. None were interested in licensing the technology. He with his backer, Harry Cutts, formed the Harwood Self-Winding Watch Company which commissioned the Swiss firms Fortis and A. Schild to make the watches. They went on sale in 1928 and 30,000 were made before the company went under due to the Depression in 1931.

The watch companies began to take an interest in coming up with solutions to the self-winding watch themselves. Fortis and A. Schild collaborated in a lever winding mechanism while Leon Hatot, a French company, thought of the entire movement gliding back and forth within a frame (this was made for them by a Swiss company). The concept which was most successful was by the Rolex Watch Company in 1930. Instead of a weight swinging back and forth, the weight swung in a complete circle as a 'rotor' movement. Sold as the Rolex Oyster Perpetual, this was the beginning of Rolex's fortunes as perhaps the most successful traditional watchmaker.

FIG.1.

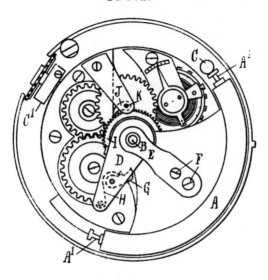

FIG. 2.

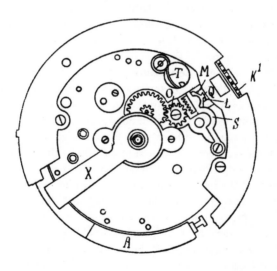

FIG.3.

FIG. 4.

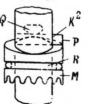

Synthetic insulin

Drug regulating the supply of sugar to the bloodstream
Fredrick Grant Banting, Charles Herbert Best and James Bertram Collip for the
University of Toronto, Canada
Filed 15 January 1923 and published as **CA 234336, GB 203778** *and* **US 1469994**

Frederick Banting was born on an Ontario farm in 1891. He took a degree in medicine, and began working in general practice. He became interested in the problem of controlling the level of sugar in the bloodstream. Too high a level means diabetes, which leads to terrible side effects as the weakest organs are attacked. Too low a level, ironically, is considerably rarer. In 1889 researchers had realised that the pancreas controlled the sugar level. Some sort of chemical messenger—insulin, as it turned out—was coming from the 'islet tissue' in the pancreas and doing this work. Identifying it was important, as then some way of helping to regulate the level if it were too low could (hopefully) be worked out. Perhaps a synthetic product could be made to mimic the same effect.

Banting was not a trained researcher but decided anyway in 1920 to work on the problem. He was not aware of the huge amount of literature that had been written on preparing an active ingredient, and afterwards admitted that had he done so he might never have started. Banting found that insulin is made inactive by ferments while it remains in the pancreatic gland. By combining this fact with the known observation that tying off the duct connecting the main pancreatic gland to the intestine would atropy the main gland intact, it was possible to extract usable insulin. However, it took months for the pancreas to degenerate, and very little could be obtained. With his colleague Best, Banting searched for a chemical method to extract insulin. It was found that a low temperature and moderate concentrations of ethyl and methyl alcohol would slow up the ferment so that insulin could be extracted. These experiments were tried out on dogs and oxen.

On 11 January 1922 insulin was first administered to a patient at Toronto General Hospital (he recovered). Production was run by the Connaught Laboratories, a unit at the university. There Best in collaboration with two others improved techniques, later with the firm Eli Lilly helping. Banting was awarded the Nobel Prize for Medicine in 1923. He was so annoyed that Best did not get credit that he gave Best half of his prize money. Dr Hagedorn of Copenhagen in 1933–35 developed protamine insulin, derived from fish glands. This prolonged the effect, so that instead of administering many small doses, a day's supply could be given in a single injection. The patent was published in Britain as GB 456101. Banting was knighted in 1934. While on war work, he was killed in an air crash in Newfoundland in 1941.

PATENT SPECIFICATION

Application Date: June 13, 1922. No. 16,360/22.

203,778

Complete Accepted: Sept. 13, 1923.

COMPLETE SPECIFICATION.

A Method of Preparing Extracts of Pancreas, suitable for Administration to the Human Subject.

I, VICTOR FALLON FEENY, of 73A, Queen Victoria Street, London, E.C. 4, a British subject, do hereby declare the nature of this invention (a communica-
5 tion to me from Frederick Grant Banting, James Bertram Collip and Charles Herbert Best, all of the University of Toronto, City of Toronto, County of York, Province of Ontario,
10 Dominion of Canada, and all British subjects), and in what manner the same is to be performed, to be particularly described and ascertained in and by the following statement:—
15 This invention relates to a method of preparing, for use in the treatment of diabetes, a pancreas extract which contains the anti-diabetic principle or hormone.
20 Many years ago it was proposed to prepare an injection for the treatment of diabetes, from the pancreatic gland taken from an animal, in which process the gland was left to self-digestion: it
25 was stated that the albuminous bodies were precipitated by alcohol and the filtrate was finally evaporated *in vacuo*. In that process there was a predisposition of the pancreas by selecting the condi-
30 tions of its removal from a narcotised animal. In another process it was proposed to obtain a digestive medical compound comprising pancreatin, which compound was formed by taking sweet-
35 breads, chopping them up finely, digesting them with alcohol, and pressing drying and pulverising them.
The communicators have ascertained by experiment that for the pancreas
40 extract to be suitable for the treatment of diabetes by way of intravenous or subcutaneous administration, it is essential that the anti-diabetic principle or hormone should be purified from the
45 enzymes, proteins, lipoids and salts which have been extracted from the pancreas.

The communicators have reason to believe that any step which involves self-digestion of the finely divided pancreas, would allow the enzymes to destroy or to 50 affect injuriously the active principle or hormone.
According to the present invention the pancrease is first treated with a solvent, such as alcohol, which inhibits the 55 deleterious action of the enzymes on the hormone; the proteins and salts are then removed as far as possible from the solution by precipitation, and finally separating by precipitation any substances which 60 may be still present with the hormone. A suitable method of carrying out this invention consists in finely dividing the pancreas and mixing it with alcohol, straining or filtering the mixture to 65 separate inert gland tissue from the substances dissolved in the alcohol, treating the filtrate with additional alcohol, removing the precipitate by filtration and concentrating the filtrate, treating the 70 latter with ether, eliminating the ether, adding alcohol, centrifuging the mixture to cause it to form into layers so that the uppermost one consists of alcohol holding the hormone in solution, the lower 75 layer or layers consisting of flocculent protein, salt solution and salt crystals, subsequently removing the uppermost layer and treating the same with several volumes of 95% ethyl alcohol to separate 80 by precipitation any substances which may be still present with the hormone, the alcohol being thereafter distilled off and the hormone dissolved in distilled water, and in concentrating the aqueous 85 solution, sterilising it and adding a preservative.
It is well known that carbohydrate, such as the starches, taken in the food, is converted into simple sugars, such as 90 glucose. In this form it is absorbed by the intestine and carried to the liver

[*Price 1/-*]

Television

John Logie Baird, The Lodge, Helensburgh, Dunbartonshire, Scotland and Wilfred Day, Highgate, London, England
Filed 26 July 1923 and published as **GB 222604**

John Logie Baird's patent for a 'System of transmitting views, portraits, and scenes by telegraphy or wireless telegraphy' was the first for a working television. However, it was a mechanical rather than an electronic version. It was based on DE 30105, an 1884 patent by Paul Nipkow of Berlin. This 'electric telescope' involved two identical spinning disks, one in the transmitter and one in the receiver. Each disk had 24 square holes and photocells transmitted the image. It is the same idea as the flicker books that schoolchildren make, and the same concept of persistence of vision means that still images are transmitted fast enough to give the impression of motion. This is the principle of television (and of motion pictures).

Baird was born in 1888. An unsuccessful businessman, he was impoverished during his early research and the first television sets were made from anything to hand—tea chests, biscuit boxes, darning needles and so on. The spinning disks were made of cardboard. The images had 30 lines from top to bottom and were transmitted at 10 times a second. A tiny image of a Maltese cross was successfully transmitted at a range of few metres in 1924. The first person ever to be seen on television was William Taynton, an office boy, on 2 October 1925. At first no image was seen as he had moved, scared of the heat from the primitive camera. He had to be bribed to be 'televised'. By 1926 Baird was becoming successful with demonstrations at Selfridge's (for £20 a week) and the Royal Institution.

In 1928 Baird began to make regular transmissions and was even experimenting with colour. His mechanical method was, however, a technical dead-end. Although he did improve the quality it was never going to be particularly good. The flickering images gave viewers a headache, and the filming involved bathing the scene in intense light which was exhausting for the cast. Research by others involved an electronic method of creating and receiving the images. Many were active in the field, including the Russian-American Vladimir Zworykin with his US 2141059 (filed in 1923 but only published in 1938) and Philo Farnsworth with US 1773980, published in 1930. Britain's EMI worked on the Zworykin ideas under licence.

In 1936 the British Broadcasting Corporation, which used radio, decided to broadcast by television for the first time. The two systems, EMI and Baird, were set up at Alexandra Palace in north London and were told to compete in alternate weeks to see which gave the better results. The electronic variety was so superior that after 3 months Baird was told to close down his system. Baird died in 1946 at Bexhill, Sussex.

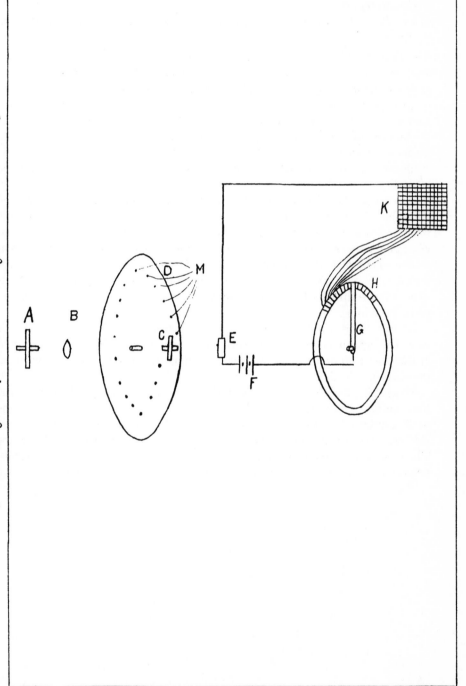

The Theremin

Electronic music producer
Lev Sergeivitch Termen, Leningrad, Soviet Union for M. J. Goldberg und Sohne,
Berlin, Germany
Filed 8 December 1924 in Germany and published as DE 443536 *and* US 1661058

The Theremin was the first electronic instrument. Termen was a Russian cellist and electronic engineer, who was born in St Petersburg in 1896. He worked on the idea of producing music by using the heterodyning effect. This is when two high radio frequency sound waves of similar but varying frequency combine and create a lower audible frequency. The sound produced is equal to the difference. The presence of the human body affects the sounds produced. Far from seeing this as a problem, Termen intended to use this so that music could be produced by 'playing' the instrument by waving his hands over it. With practice, complicated music could be 'played'.

Termen developed his first model in 1917, using vacuum tubes, and called it the Theremin. There was a foot pedal to control the volume and a switch to alter the pitch. It was first demonstrated in public in 1920 and Lenin was so impressed that he asked for lessons. Six hundred were built and, at Lenin's command, many went on tour round the USSR. In 1927 Termen left for the United States. He set up what for the time was a spectacular electronic studio in New York, hoping to win donations from wealthy backers. The Radio Corporation of America (RCA) began to make Theremins.

In 1938, Termen and his wife, a singer, were kidnapped by the NKVD from their New York apartment and were spirited away to the USSR. There he was accused of anti-Soviet propaganda and was sent to a labour camp. While there he invented the electronic 'bug'.

Meanwhile, the Theremin was acquiring a certain popularity. The eerie music that you hear on science fiction films from the 1950s is from one—and so is the backing electronic music on the Beach Boys' hit, *Good vibrations*. A transistor version was later made but Robert Moog's synthesiser eventually replaced the Theremin as the method to generate electronic music. Termen was later rehabilitated and died in 1993, having been responsible for many other patents.

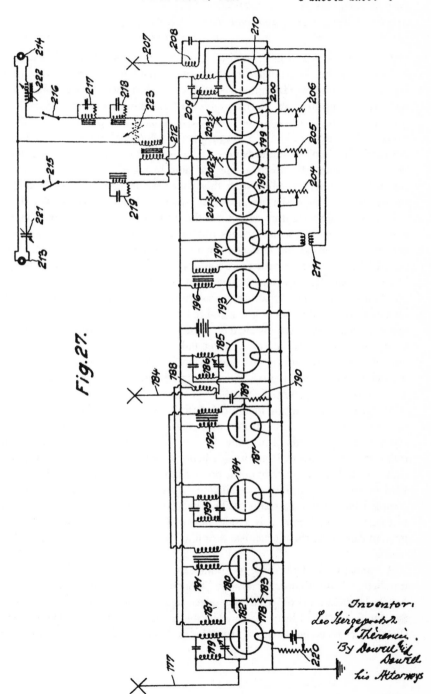

Fig. 27.

Inventor:
Leo Sergeivitch
Thèremin.
By Dowell & Dowell
his Attorneys

Traffic lights

Signals controlling traffic at road junctions
Garrett Augustus Morgan, Cleveland, Ohio
Filed 27 February 1922 and published as US 1475024

Garrett Morgan was born in Kentucky, the son of former slaves. He had little formal education but loved experimenting with gadgets while making a living repairing sewing machines. Having moved to Cleveland in 1895, he was able to open his own repair shop in 1907. He later added a flourishing tailoring business. In 1914 Morgan patented two safety hoods for use in smoky atmospheres, which were published as US 1090936 and US 1113675 for the National Safety Device Company. Although they look in the drawings like something out of a science-fiction movie they were apparently effective. He was able to use them to rescue several men trapped in a tunnel on 25 July 1916; this made national news. Orders flooded in, and the equipment was used as gas masks by American troops in World War I.

His second important invention was the traffic light. More and more motor cars were sharing the road with carriages and bicycles. One day Morgan saw an accident between a motor car and a horse-drawn carriage which involved a little girl being seriously injured. At the time there was a 'stop/ go' mechanism, manually operated by a policeman. Besides being expensive in manpower, these did not allow for complete halts, and assumed that traffic was always moving in one direction or another. He came up with the idea of green and red signals combined with a warning buzzer and managed to get them installed on 5 August 1914 by the American Traffic Light Company at the corner of 105th Street and Euclid Avenue in Cleveland. This was not patented. There are supposed to have been versions by other inventors, including one in New York in 1918 with red, green and amber lights, but these do not seem to have been patented either.

Morgan's invention provided for electrically powered cranks turning the arms so that they could halt traffic or indicate a right of way. He still wanted the 'customary' stop and go words of his time. Figure 1 shows the invention with on-coming traffic at stop and traffic from the side being at go (this is indicated at the end of the right arm and at the top right of the post). Figure 2 showing the two side arms swivelled up to stop all traffic, since all sides of the arms state 'stop', to provide for a complete halt of all road traffic to allow pedestrians to cross. Figure 3 is a side view of Figure 1, and Figure 4 is a side view of Figure 2. He provided for means to prevent damage to the mechanism dropping down with a dashpot for absorbing the inertia.

A manual traffic light came to London in 1926, with an automatic system being installed at Wolverhampton in 1927. Morgan's invention was widely used across North America until replaced by modern lights. He eventually sold his rights for $40,000 to General Electric, a considerable amount for the time. He died in 1963, shortly after receiving an award from the American government for road safety.

Nov. 20, 1923.

G. A. MORGAN

1,475,024

TRAFFIC SIGNAL

Filed Feb. 27. 1922

2 Sheets—Sheet 1

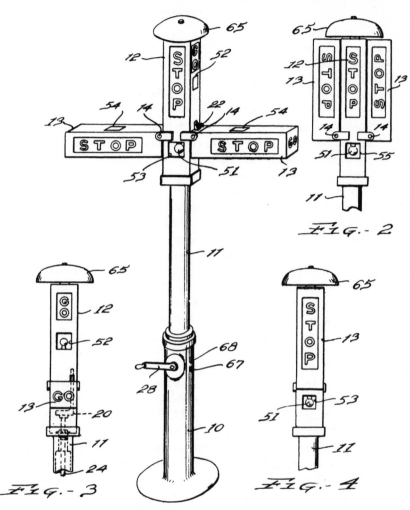

FIG.-2

FIG.-3

FIG.-4

FIG.-1

INVENTOR
Garrett A. Morgan,
By Baker Macklin,
ATTORNEYS

Transparent adhesive tape

Richard Drew, St Paul, Minnesota for Minnesota Mining and Manufacturing Company
Filed 30 May 1928 and published as US 1760820 *and* GB 312610

Richard Drew was born in 1886. He joined the Minnesota Mining and Manufacturing Company (later 3M) in 1923, when it was a small company making sandpaper. He had been experimenting with a new kind of sandpaper, and was visiting garages to ask them to try it out, when he heard that their employees were having problems when painting cars with two colours. When they used heavy masking tape to protect the already painted part, the paint often came away when the tape was removed. Drew realised that if you took away the abrasives from sandpaper you already had the backing paper and the adhesive for such a product. He went away and developed a masking tape consisting of a 5-centimetre wide tan paper strip coated with a light, pressure-sensitive adhesive.

The prototype had adhesive along its edges but not in the middle. When first tried out at a garage the tape fell off the car. The painter said to Drew, 'Take this tape back to those Scotch bosses of yours and tell them to put more adhesive on it!' By 'Scotch', of course, he meant stingy. Drew accordingly improved the adhesive tape. At the same time he was investigating a watertight backing that could be used as a moisture-resistant sealant for the entire contents of a refrigeration car on the railways. Cellophane had recently been invented by Du Pont, and was watertight. However, it was also heat-sensitive and adhesive-resistant. Drew began experimenting to make cellophane more practical for his purpose. In the meantime many food packaging firms had heard about his masking tape for cars and were asking about waterproof sealants for food sold in stores.

Drew was told by his boss to stop working on the project and to go back to sandpaper work, but Drew ignored him. He also fiddled his expense limit of $100 without authorisation by spending $99 at a time. He finally came up with an adhesive made from a mixture of rubber, oils and resin which formed a coat on a cellophane backing. He remembered the jibe about 'Scotch', and named it Scotch® Tape.

At first it was sold to food producers sealing individual cellophane bags of food, but soon Du Pont came up with a method of heat-sealing such bags. The emphasis switched to selling the product direct to consumers. The poverty of the Great Depression meant that many uses were found for it, from mending clothes to preserving cracked eggs. The familiar dispenser was invented a few years later by John Borden, one of the company's sales managers, and plaid was adopted for use on the products from the 1940s to accord with the Scottish theme. The company switched permanently from abrasives to sealants and coatings, and continued to develop the tape. It was so popular by World War II that the company ran advertisements apologising for shortages. Richard Drew died in 1956.

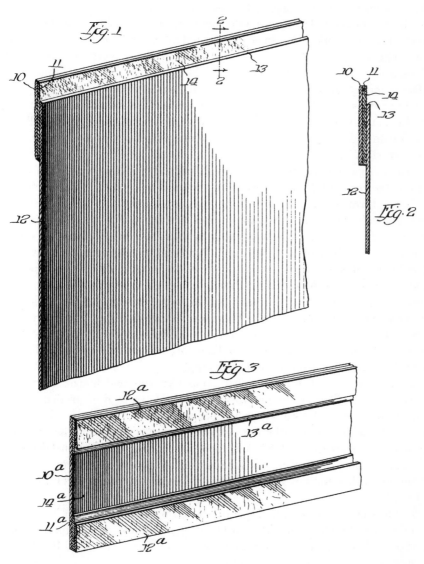

1930–1939

W. H. AUDEN called the 1930s 'a low dishonest decade'. Economic recovery after World War I fractured in 1929 and the following decade was dominated by record levels of unemployment in many countries and by the drift towards another world war. The League of Nations was too weak to resist aggression by Hitler's Germany in Austria and Czechoslovakia, by Mussolini's Italy in Ethiopia and Albania and by Japan in Manchuria and (with huge loss of life) in China. These fascist regimes looked increasingly towards greater 'living space' and armaments production as solutions to economic hardship. Only with the invasion of Poland did Britain and France call a halt to Hitler's unchecked occupation of other countries: World War II began.

Gandhi's peaceful agitation which would lead to India's independence gathered pace in the 1930s while nationalist and communist movements fought each other in China, turning Mao Zedong's Long March into legend. In the Spanish Civil War Franco's nationalist and military forces triumphed over the republican government after a struggle which was seen by many to symbolise conflict between democracy and dictatorship. Germany's Nazi regime removed civil rights, outlawed political opposition and trade unionism and persecuted the Jewish population and other minorities. Stalin reinforced his ruthless supremacy in the Soviet Union through economic 5-year plans and by the terrorising purges of politicians, military leaders and intellectuals. There was a huge death toll and deprivation, especially among the wealthier peasantry (the Kulaks), during Soviet agricultural collectivisation that compressed 25 million single farms into 250,000 collectives with the aim of producing a better food supply.

There were bank failures in Austria and Germany and hunger marches in Britain, and the Depression was worsened in the USA by drought that brought poverty to farmers and caused 70,000 people to leave the 'dust bowl' states in search of food and wages. But the economic reconstruction of Roosevelt's New Deal programmes in the USA led, for example, the new National Recovery Administration to legislate on far-reaching employment reforms and the Tennessee Valley Authority to provide electricity and help for agriculture among other purposes.

Landmarks in civil engineering were reached with the opening of New York's Empire State building (then the world's tallest building), the Sydney Harbour Bridge in Australia and the longest bridge of all, San Francisco's Bay Bridge. Motorways (and autobahns in Germany) were increasingly encountered by motorists and both flight and motor technology developed rapidly. The 1930s saw the first regular trans-Pacific passenger and mail flights, a record-breaking 'round the world' flight led by Howard Hughes, Amy Johnson's trans-continental air crossings and Piccard's pioneering ascent to the stratosphere via balloon. The loss of the *Hindenburg*, though, dealt a fatal blow to airship production. Cobb broke the car land speed record and a 'people's car', the Volkswagen designed by Ferdinand Porsche, was announced in Germany; its 'Beetle' models produced post-war, and some until the 1980s, would eventually outsell the 15 million cars of Ford's Model T.

Much invention for home and work flourished even in the decade's disturbed times. Fluorescent lighting, the photoflash bulb, 35 mm Kodachrome film and the photocopier (which would link a chain between the earlier paperclip and later computer in revolutionising office work) were all developed and the first synthetic detergent (by ICI) and DDT insecticide were produced. In science, technology and medicine King isolated vitamin C, Richter invented a scale to measure earthquakes, Chadwick discovered the neutron, Jansky cosmic radio waves and Hahn the process of nuclear fission. Cockcroft became the first to split the atom by mechanical means and Lawrence built an 'atom smasher' (the cyclotron) which would be crucial to the atom bomb's production.

While science and technology pushed forward, William Plomer's poem set in 1939 captured some of the emotional and political recklessness, inequality and tragedy present in the 20 years' armistice between the two world wars:

> . . . the twenties and the thirties were not otherwise designed
> Than other times when blind men into ditches led the blind,
> When the rich mouse ate the cheese and the poor mouse got the rind,
> And man, the self-destroyer, was not lucid in his mind . . .

The Anglepoise® adjustable lamp

Adjustable table lamp
George Carwardine, Bath, Somerset, England
Filed 4 July 1932 and published as **GB 404615**

George Carwardine had had a career as an engineer in a small English car company, and took out many patents, especially for suspension systems. This gave him a great deal of knowledge about springs and the problems of movement. He began to take an interest in an apparatus which could move easily through three planes but would be rigid when left free, although he had no clear idea of a useful product. Years later he realised that the concept would be useful for focusing light on, say, a table to help reading.

He worked on making the concept as versatile as an arm. At first there were two springs connecting each arm, but fingers could be caught in the mechanism. This was changed so that three tight springs at the base controlled the movement. At first he called it the Equipoise, but this was rejected by the Patent Office as a trade mark since this was considered to be an existing word. Hence Anglepoise® was registered instead in 1947 (and in 1950 in the United States).

Carwardine licensed production to Herbert Terry & Sons, partly because they were experts in coiled springs. Although no effort was made to add beauty to the product its clean and functional lines have won the Anglepoise® a reputation as a design classic. After World War II the company gave away the rights to manufacturing in continental Europe and the United States to Jac Jacobsen, a Norwegian entrepreneur. Jacobsen launched the product in the United States as the Luxo in 1951 and achieved great success. He would tell the Terry family at trade fairs how kind they had been to him.

The product continues to be used in various modifications. The Queen uses Anglepoise® lamps, as do the scientists in Q's laboratory in the James Bond films. Harrison Ford and Sean Connery use one to help decipher codes in the film *Indiana Jones: the last crusade*. More seriously, the lamps are frequently used in hospitals and for other special purposes. One of the early uses was to provide light for the navigators of Wellington bombers in World War II.

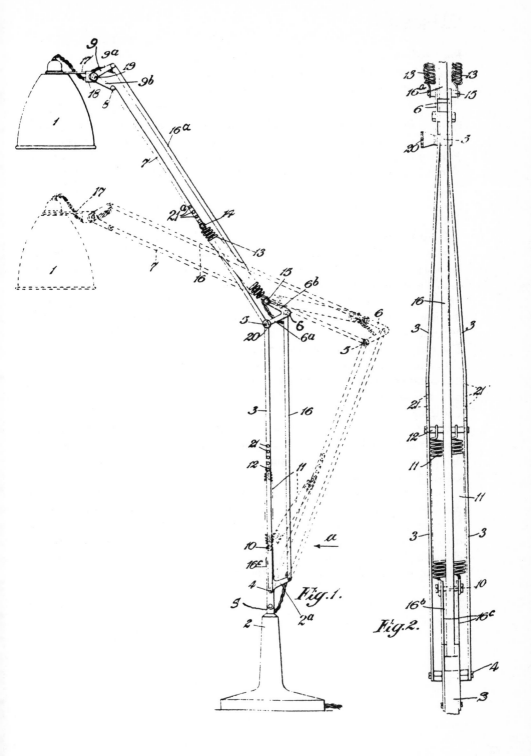

Fig.1.

Fig.2.

Catseyes®

Road reflectors
Percy Shaw, Halifax, Yorkshire, England
Filed 30 May 1935 and published as **GB 457536**

Percy Shaw, born in 1890, was a road mender. During the course of his work, and while driving, he realised that winding, unlit roads were dangerous at night, even with headlights. On one occasion he was driving cautiously on a dark and foggy night when he realised that he could see two small green lights, close to each other. They were the light of his headlights reflected in the eyes of a cat. He wondered if he could make a cheap and simple device that would do the same job when embedded in the road.

Catseyes® are two small metal marbles placed close to each other within a hard rubber casing. This is set into the road at intervals between the lanes of traffic and forms a small hump. As the motorist travels along, the road lights up well ahead because the beams of the headlights are reflected by the Catseyes®. Perhaps the most ingenious aspect is the cleaning of the marbles. When it rains some water collects in the depression holding the marbles. When the occasional car goes over the hump it depresses it so that the water is flushed out, cleaning the marbles as it does so. Therefore there is minimum maintenance.

In 1935 Shaw formed Reflecting Roadstuds Ltd to make Catseyes®, a company that continues in operation to the present day. Its factory was set up at Boothtown on a site next to his own house. During the late 1940s a junior minister, James Callaghan, authorised the purchase of millions of Catseyes® for installation in the roads. He went on to become Prime Minister during 1976–79 but felt that this was one of his most important achievements. Catseyes® continue to be found on many minor British roads. According to legend Shaw received a farthing for each Catseye® that was installed.

Shaw patented other ideas to do with road markings or signs, including improvements to the Catseye® concept. Various trade marks using the name or a symbolic appearance of a cat's eye were also registered from 1938. He enjoyed being driven round Yorkshire in a Rolls Royce, stopping occasionally to purchase fish and chips which he ate out of newspapers in the back of the car. Shaw was made an OBE (member of the Order of the British Empire) in 1965. He died in 1976.

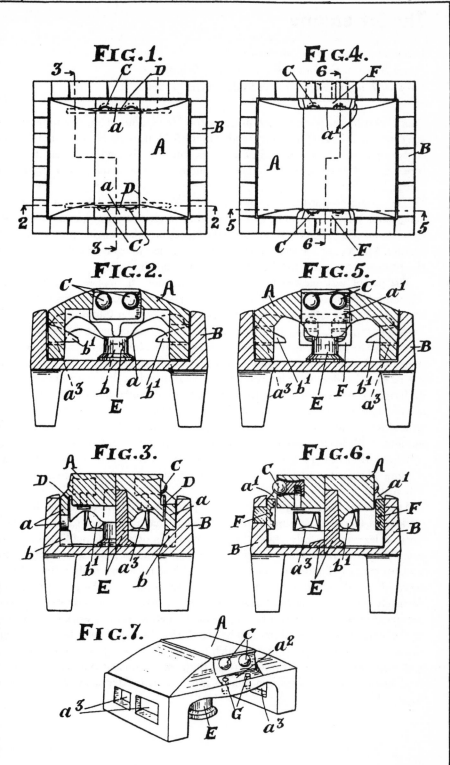

FIG.1.

FIG.4.

FIG.2.

FIG.5.

FIG.3.

FIG.6.

FIG.7.

Malby & Sons, Photo-Lith

The jet engine

Using jet propulsion for forward thrust
Frank Whittle, Coventry, Warwickshire, England
Filed 16 January 1930 and published as **GB 347206**

Frank Whittle was born in 1906 in Coventry. He became an aircraft apprentice in 1923 and was selected for officer training in the RAF in 1926. He suggested the concepts of rocket propulsion, and of using gas turbines, but not together, in a term paper in 1928. Whittle later thought of combining the two concepts in a single engine and filed his patent. Figure 1 shows the energy cycle and Figure 2 the 'device' (not in correct proportions). A compressor 1 takes in air and delivers it to a lagged combustion chamber 10. The gases expand through a turbine 13 and are expelled through nozzles 17. Problems included the blades breaking, engine surges and overheating. Much practical experimentation was needed to overcome these problems. His patents are unusual for the clarity of their English.

The patent was allowed to lapse from protection in 1935 for lack of money. Whittle had more or less given up when in 1935 two ex-RAF officers encouraged him while he was doing a sponsored engineering degree at Cambridge. In 1936 he formed a company, Power Jets, to exploit his ideas, backed by an investment bank. From 1937 to 1946 the RAF placed him on special duty which meant that he could work full-time on his ideas, although it was not until 1939 that the Air Ministry admitted that the ideas were interesting. The engine had its first test run in 1937 but performance was erratic and better materials needed to be found. Whittle himself solved a hitherto unknown problem with turbine blades, although turbines had been around for decades: it had not been realised that the air flow would change as the number of blades increased, and redesigning would be needed. An experimental jet fighter, the Gloster E28/39, was built and flew for the first time on 15 May 1941. A more advanced model, the Gloster Meteor, flew in combat in 1944, the only Allied plane to do so.

The Gloster was in fact not the first jet plane to fly. Hans von Ohain had been working in Germany on similar lines since 1933. Each was unaware of the other's work, at least in the early years, although Whittle's patent was published in 1931. Von Ohain's Heinkel He 178 flew for the first time on 27 August 1939. Hitler watched an early flight and commented 'Why is it necessary to fly faster than the speed of sound?' Other German scientists were working on the problem as well. Strangely enough, no aircraft manufacturer was involved in the early work in either country. Meanwhile, General Electric in the United States was working from 1941 on the basis of plans passed to them by the British. In 1944 Power Jets was nationalised by the British government, which bought out Power Jets' many patents for £100,000, and knighted Whittle, in 1948. Both von Ohain and Whittle emigrated to the United States, von Ohain to Florida and Whittle to Maryland. Whittle died in 1996 and von Ohain in 1998.

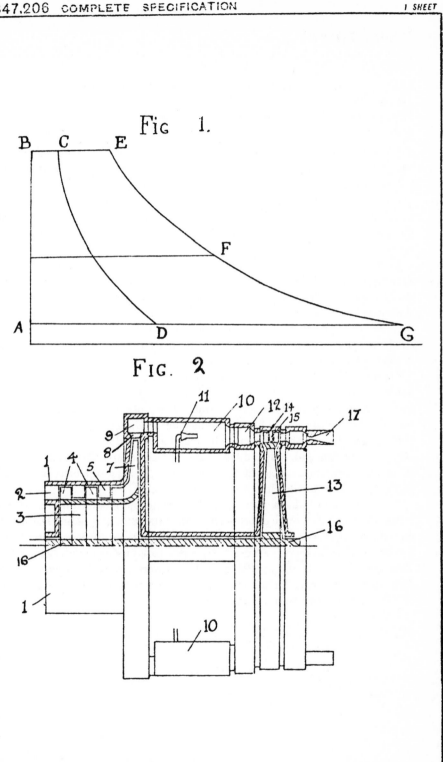

FIG. 1.

FIG. 2

Monopoly®

Board game
Charles Darrow, Philadelphia, Pennsylvania for Parker Brothers, Salem,
Massachusetts
Filed 31 August 1935 and published as US 2026082

The usual account is that Charles Darrow was a salesman who thought up the game while unemployed in the Depression. He had spent many holidays in Atlantic City, New Jersey, and the American version therefore had squares named after streets in that resort. He painted the board on a tablecloth and built tiny houses, hotels and so on out of scrap pieces. His unemployed friends gathered to play this fantasy money game and asked if they could have sets too, so he began to sell sets for $4 each.

Darrow wrote to Parker Brothers asking if they were interested in selling it but they replied that there were 52 fundamental errors in the game, including that it took too long to play, it was too complicated, and there was no clear goal (the last, at least, is surely wrong). He then approached department stores, and word got to the wife of the president of Parker Brothers, Robert Barton. He listened to his wife and bought a set, and later bought the rights. The trade mark was registered in 1935 in the United States and in 1952 in Britain.

Darrow became the first games inventor to become a millionaire. A plaque in his honour is in Atlantic City, near Park Place. Foreign versions feature London or Paris streets and so on. That is the normal version of how the game was invented, but it has been challenged. Other accounts point to Lizzie Magie of Maryland with her US 748626, published in 1905. This board game had no street names but has a close resemblance, with rent and sale prices on each square. She meant the game to be an attack on capitalism as she was against rent, being a supporter of the radical Henry George. She called it 'The landlord's game'.

Dan Layman of Indiana got to hear of the game and modified it in a version he called 'Finance'. By a chain of people hearing of the game and adding features it acquired Atlantic City street names by a local resident and then was introduced to Darrow. Parker Brothers is alleged to have purchased rights to these games to prevent litigation. In about 1974 Ralph Ansbach reverted to earlier versions with his Anti-Monopoly® game. He was charged with trade mark infringement. He won his case in the Supreme Court in 1984.

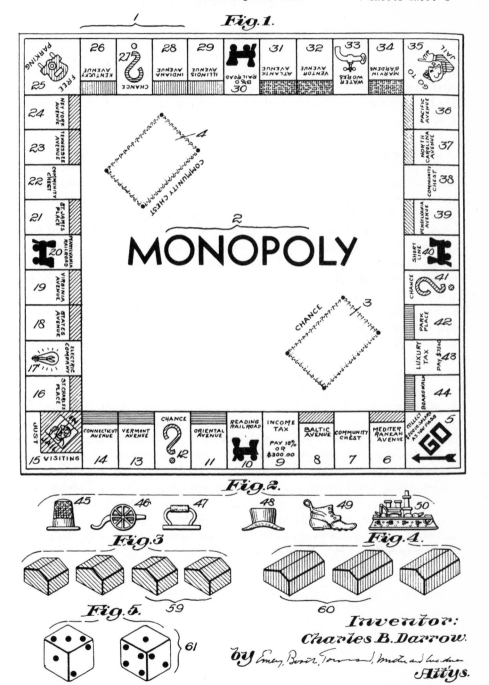

Fig.1.

MONOPOLY

Fig.2.

Fig.3.

Fig.4.

Fig.5.

Inventor:
Charles B. Darrow.
by Emery, Booth, Townsend, Weston and Weston
Att'ys.

Nylon

Synthetic fibre
Wallace Hume Carothers, Wilmington, Delaware for E. I. du Pont de
Nemours
Filed 9 April 1937 and published as US 2130948 *and* GB 461236–7

Du Pont called nylon 'the first man-made organic textile fiber wholly from new materials from the animal kingdom'. Wallace Carothers was a basic researcher at the Du Pont laboratories. This means that he was not actively developing a particular type of product but rather finding out as much as he could and seeing if anything developed.

He was asked to investigate polymers, a new area of research. These are long strings of molecules that are found in rubber and silk. If the company could understand how polymers were constructed, it might be able to create fibres from raw materials such as coal or petroleum rather than from plants or silkworms. Carothers decided to build synthetic polymers so that the process that keeps them as strings could be understood. By 1930 he and his team of researchers had decided that polymers were special just in that they were longer than other structures. By 1933 things looked dark. Although they had a synthetic polymer that looked like silk, it could not be spun into fibres as it melted too easily, and heat was needed in the process. This polyamide was, in fact, nylon.

The researchers turned to another type of plastic, polyesters. Julian Hill, a colleague, tried putting a glass rod into a beaker full of polyester, sticking a piece firmly on, and removing it. The result was long strands like a spider's web as he drew it out. This is called 'cold-drawing'. The story is that the staff waited until Carothers, their boss, had left the building to try cold-drawing in the hallway. To their amazement they were able to stretch a piece of polyester the entire length of the hallway. It was then realised that the act of cold-drawing made the substance stronger by making the molecules line up instead of being in a jumble. The same process was then tried on the forgotten nylon, and it was found that it was stronger than silk.

Du Pont decided to sell nylon in the hosiery market, as stockings often 'ran'. This effort was so successful that they came to be called 'nylons' and were a valuable commodity in World War II. Carothers himself fell into a depression. This has been blamed on fears that Du Pont did not want to make the product a success, but he was prone to depressions, and a beloved sister also died suddenly. On 29 April 1937 Carothers took cyanide in a Philadelphia hotel room, 3 weeks after the patent application had been filed. He was 41.

2,130,948

SYNTHETIC FIBER

Wallace Hume Carothers, Wilmington, Del., assignor to E. I. du Pont de Nemours & Company, Wilmington, Del., a corporation of Delaware

No Drawing. Application April 9, 1937,
Serial No. 136,031

56 Claims. (Cl. 18—54)

This invention relates to new compositions of matter, and more particularly to synthetic linear condensation polyamides and to filaments, fibers, yarns, fabrics, and the like prepared therefrom.

The present application is a continuation-in-part of my application Serial Number 91,617, filed July 20, 1936, which is a continuation-in-part of application Serial Number 74,811, filed April 16, 1936, which is a continuation-in-part of abandoned application Serial Number 34,477, filed August 2, 1935, which in turn is a continuation-in-part of application Serial Number 181, filed January 2, 1935; and of U. S. Patent 2,071,251, filed March 14, 1933; and of U. S. Patent 2,071,250, filed July 3, 1931.

Products obtained by the mutual reaction of certain dibasic carboxylic acids and certain organic diamines have in the past been described by various investigators. For the most part, these products have been cyclic amides of low molecular weight. In a few cases they have been supposed to be polymeric, but they have been either of low molecular weight or completely infusible and insoluble. In all cases, they have been devoid of any known utility. These statements may be illustrated by the following citations: Ann. 232, 227 (1886); Ber. 46, 2504 (1913); Ber. 5, 247 (1872); Ber. 17, 137 (1884); Ber. 27 R, 403 (1894); Ann. 347, 17 (1906); Ann. 392, 92 (1912); J. A. C. S. 47, 2614 (1925). Insofar as I am aware, the prior art on synthetic polyamide fibers, and on polyamides capable of being drawn into useful fibers, is non-existent.

This invention has as an object the preparation of new and valuable compositions of matter, particularly synthetic fiber-forming materials. Another object is the preparation of filaments, fibers, and ribbons from these materials. A further object is the manufacture of yarns, fabrics, and the like from said filaments. Other objects will become apparent as the description proceeds.

The first of these objects is accomplished by reacting together a primary or secondary diamine (described comprehensively as a diamine having at least one hydrogen attached to each nitrogen) and either a dicarboxylic acid or an amide-forming derivative of a dibasic carboxylic acid until a product is formed which can be drawn into a continuous oriented filament. The second object is attained by spinning the polyamides into filaments, and preferably, subjecting the filaments to stress ("cold drawing") thereby converting them into oriented filaments or fibers.

The third of these objects is accomplished by combining the filaments into a yarn and knitting, weaving, or otherwise forming the yarn into a fabric.

The term "synthetic" is used herein to imply that the polyamides from which my filaments are prepared are built up by a wholly artificial process and not by any natural process. In other words, my original reactants are monomeric or relatively low molecular weight substances.

The term "linear" as used herein implies only those polyamides obtainable from bifunctional reactants. The structural units of such products are linked end-to-end and in chain-like fashion. The term is intended to exclude three-dimensional polymeric structures, such as those that might be present in polymers derived from triamines or from tribasic acids.

The term "polyamide" is used to indicate a polymer containing a plurality of amide linkages. In the linear condensation polyamides of this invention the amide-linkages appear in the chain of atoms which make up the polymer.

The terms "fiber-forming polyamide" is used to indicate that my products are capable of being formed directly, i. e., without further polymerization treatment, into useful fibers. As will be more fully shown hereinafter, fiber-forming polyamides are highly polymerized products and for the most part exhibit crystallinity in the massive state.

The term "filament" as used herein refers to both the oriented and unoriented filaments or threads which are prepared from the polyamides regardless of whether the filaments or threads are long (continuous) or short (staple), large or small, while the term "fiber" will refer more specifically to the oriented filaments or threads whether long or short, large or small.

The expression "dibasic carboxylic acid" is used to include carbonic acid and dicarboxylic acids. By "amide-forming derivatives of dibasic carboxylic acids" I mean those materials such as anhydrides, amides, acid halides, half esters, and diesters, which are known to form amides when reacted with a primary or secondary amine.

The following discussion will make clear the nature of the products from which my filaments and fibers are prepared, and the meaning of the above and other terms used hereinafter. If a dicarboxylic acid and a diamine are heated together under such conditions as to permit amide formation, it can readily be seen that the

The parking meter

Device charging for car parking
Carl Magee for the Dual Parking Meter Company, both Oklahoma City, Oklahoma
Filed 13 May 1935 and published as US 2118318

Carl Magee was the editor of an Oklahoma City newspaper. There was concern about the difficulties of parking on the busy city streets as those who worked there were parking all day, preventing others from parking while shopping. At the time the usual way of fining car drivers, if they parked their cars for too long, was for a police officer to mark the tyres with chalk, and then see if the marks were still there a few hours later. This method obviously used a great deal of manpower, apart from the risk of a driver removing the mark (or of delinquents marking cars for fun).

Magee was appointed to the city's Chamber of Commerce Traffic Committee to see if the problem could be solved. Rightly or wrongly, he seems to have decided to make money by forming his own company to manufacture a solution. In December 1932 he filed for a patent for a 'parking meter' which was published as US 2039544. This was a primitive device looking like a rectangular box perched at the top of a pole. Rights in it were assigned to Magee's Dual Parking Meter Company: dual because it had the twin roles of raising revenue for the city and of controlling congestion. The company is still operating, but it is now called POM Inc.

The classic parking meter was improved with his US 2118318. He states that when parking, 'It is desirous an incidental charge be made upon a time basis'. Figure 1's modern lines are the same as the first to be installed. Five pages are taken up with explaining how the mechanism works to measure time elapsed, which is shown in Figure 2 and in much more detail in other figures (not shown here). A spring operates this together with two cogs, having been powered by the user turning a handle after inserting coins. Since the hour indicator does not move much, little power is in fact needed. The principles of the machine are commonly used to this day.

The city authorities placed an initial order for 150 units and the first meter was installed on 16 July 1935. It took a month for the first driver to be fined by parking beyond his stay, a Rev. North. Production continued at Oklahoma City and at Tulsa until 1963. Parking meters only arrived in Britain in 1958, in Mayfair, London. Meters have caused a great deal of angst and, to put it mildly, interest among the driving public. Attempts have been made to jam them so that they would not work (in London and no doubt elsewhere jammed meters meant free parking for a while until the authorities realised what was happening), as well as break-ins to get the coins. It is also necessary periodically to remove the coins as well as monitoring expired meters. More recent efforts in the field of controlling parking congestion include prepaid cards, resident-only parking and park and ride schemes. In some places there is one machine and slips of paper are taken from them and displayed within the car. One type of parking meter even photographs cars that have overstayed their welcome.

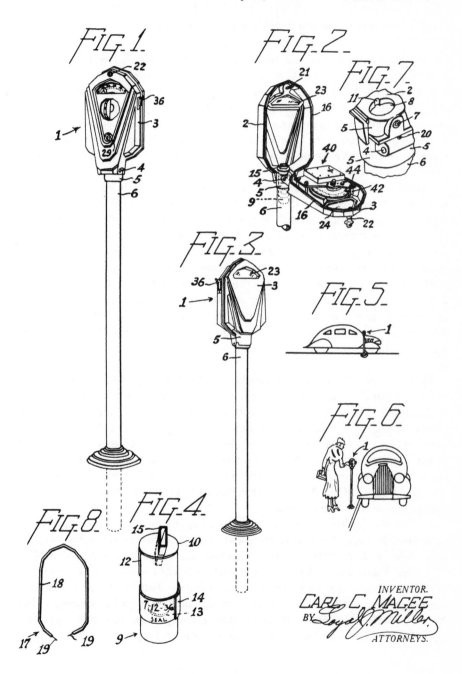

INVENTOR.
CARL C. MAGEE
BY
ATTORNEYS.

The photocopier

Electrical copier
Chester Carlson, Jackson Heights, New York City, New York
Filed 8 September 1938 and published as US 2221776

Chester Carlson's job was analysing published patents for an electrical company, P. R. Mallory. Copies were often required of the specifications, which was done by copying them by hand. It was the depression, and he couldn't find another job. He really wanted to develop inventions as he thought that was the best way to make money. He thought that there had to be a better way to make copies, and decided to carry out research at the New York Public Library. He looked through technical journals, seeking information on reproducing photographs. But this involved lots of chemicals, and took hours.

Carlson wondered if an electrical method were possible. It was already known that charged particles attach themselves to an oppositely charged surface. The trick would be to get particles to stick in a pattern identical to that of a lit-up image. A photoconductive metal plate would be charged up and an image would be projected onto it. The image would be recorded in powder, and then baked by heat onto a sheet of paper. He could figure out the concept—but could he get it to work?

He worked away in the evenings in the kitchen of his Long Island apartment, experimenting with different chemicals and substances. Legend says that the landlady's daughter was sent to complain about the smell of his experiments, only to wind up marrying him in 1934. He also somehow found the time to go to New York Law School as he thought that he would need to protect his invention, graduating in 1939.

He filed a patent for the basic concept of 'electrophotography' but had not yet made a proper dry print. Then the breakthrough occurred. A zinc plate was coated with sulphur. The plate was rubbed with a cotton cloth to create a static electricity charge. A glass slide with words written on it was held against the plate, and together they were exposed to the heat of a lamp. The slide was removed and the plate dusted with moss spores. Wax paper was pressed against the powder, heat was applied and the paper was peeled off. The words he had written were revealed: '10.-22.-38 ASTORIA'. Astoria was the suburb where he was then working, behind his mother-in-law's beauty salon.

Carlson was jubilant but could not get a company interested in developing the technique. In 1944, the Battelle Memorial Institute, a private non-profit organisation, visited P. R. Mallory to discuss other patents, and Carlson asked if they would back him. A deal was worked out involving substantial royalties for Carlson if subsequent improvements led to a marketable invention. Battelle themselves began to have problems and looked for other backers, and the Haloid Corporation (later Xerox) joined in. The first really practicable photocopier, the 914, only appeared in 1958. Carlson died in 1968, a multi-millionaire philanthropist.

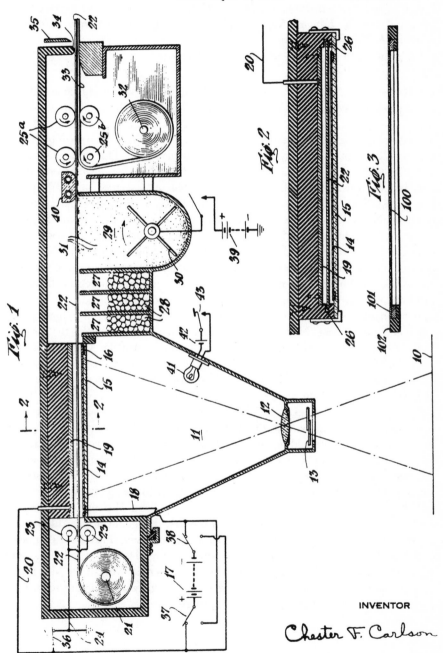

INVENTOR

Chester F. Carlson

Polythene

Polymer
Eric William Fawcett, Reginald Oswald Gibson, Michael Wilcox Perrin, John Greves Paton and Edmond George Williams for Imperial Chemical Industries, Northwich, Cheshire, England
Filed 4 February 1937 and published as **GB 471590**

Polythene (or, more properly, polyethylene) is one of the many useful chemicals discovered when chemists tried messing around with flasks. The inventors worked at a modern research laboratory built in 1928, which carried out a great deal of fundamental research in the middle of the Depression. One project was to see if high pressure would result in the production of useful materials. This was focussed on materials which normally did not react or which needed vigorous catalysts. It was already known from research at Harvard that polymers seemed to be affected by high pressure. On 24 March 1933 Gibson and Fawcett set up an experiment with crude equipment in which a mixture of ethylene and benzaldehyde was heated to 170 degrees C and put under high pressure. The apparatus was left (it was a Friday) and it was hoped that the two chemicals would merge together. When they checked on the Monday Gibson found a 'waxy solid'. This was a polymer of ethylene which could be melted and drawn into threads. Attempts to repeat the experiment all failed as either nothing happened or an explosion occurred.

In September 1935 Fawcett said at a conference that they had (once) made a solid polymer from ethylene. Nobody believed him, which was fortunate as the company had not filed a patent application. It was also fortunate that the American chemist Carl Marvel, who had made polythene by a different method three years before, had ignored it as he did not think that it had any uses. In December 1935 using improved equipment Perrin was able successfully to replicate the experiment. It later turned out that some oxygen had leaked into the flask in the initial experiment and had acted as a catalyst, while in the Perrin experiment when the pressure dropped more ethylene was pumped in and by mistake some oxygen as well, again creating the ideal conditions. In 1937 a 10.2 tonne per year pilot plant opened followed by a 102 tonne per year plant which began production on the day World War II started. Initially it was made by being gently blown up rather like soap bubbles.

Polythene is tough, easily moulded and water-repellent. Although mostly associated with transparent packaging for food, it was first thought of as an insulation material. Its first use was as insulation for radar on airplanes and ships. In 1948 it began to be used for washing up bowls and it has since been used for such products as toys, bags and cosmetic bottles. In the 1950s subsequent German work at the Max Planck Institute showed that it was possible to make polythene without high pressure but with catalysts. The British patent for this process was GB 713081.

Application Date: Feb. 4, 1936. No. 3372/36.

„   „   March 16, 1936. No. 7899/36.

„   „   Aug 21, 1936. No. 23093/36.

471,590

One Complete Specification Left: Feb. 4, 1937.

(Under Section 16 of the Patents and Designs Acts, 1907 to 1932.)

Specification Accepted: Sept. 6, 1937.

PROVISIONAL SPECIFICATION
No. 3372 A.D. 1936.

Improvements in or relating to the Polymerisation of Ethylene

We, Eric William Fawcett, Reginald Oswald Gibson, Michael Willcox Perrin, John Greves Paton and Edmond George Williams, all British Subjects, all of Winnington Hall, Northwich, in the County of Chester, and Imperial Chemical Industries Limited, a British Company, of Imperial Chemical House, Millbank, London, S.W.1, do hereby declare the nature of this invention to be as follows:—

This invention relates to the polymerisation of ethylene, with the object of producing new and useful products therefrom by effecting the polymerisation under the operating conditions hereinafter described.

It is already known that ethylene and its homologues can be polymerised to yield a mixture of liquid organic compounds by the use of elevated temperatures, e.g. 200—400° C., and of moderate superatmospheric pressures, say up to 200 atmospheres, with or without the aid of catalysts. The liquid products range in character from relatively light oils up to fairly viscous oils of the lubricating oil type.

We have now found that ethylene may be polymerised to give solid products of a rubber- or resin-like character by the use of a very high pressure, i.e. a pressure of at least 1000 atmospheres, and by the use of reaction conditions such that the heat of the reaction is removed as quickly as possible. The products obtained under these conditions appear to be true polymers, i.e. they correspond to the formula $(CH_2)_n$, and they are of high molecular weight, e.g. those obtained at 2000 atmospheres have a molecular weight of the order of 3500. The temperature requirements of the reaction, in particular, the initial temperature required for polymerisation to occur, have not yet been fully investigated, but it would appear that a moderate elevated temperature, of the order of 100—200° C.

should be employed.

It is necessary to provide for the efficient removal of the heat of reaction, otherwise it is impossible to control the reaction, which is then liable to give rise to an explosion of some violence, the end products being carbon and hydrogen. Suitable measures to achieve rapid removal of heat are the use of diluents for the ethylene (e.g. the use of industrial gases containing ethylene such as cracking still gases), and/or the use of a reaction vessel constructed of or lined with a metal or alloy of high thermal conductivity. An internal heating element of small heat capacity, disposed within a cold-walled reaction vessel, is a convenient means of initiating the reaction.

In general, increasing the pressure at which the polymerisation is effected will increase the ultimate yield of the polymer. Alternatively increasing the pressure will permit of the same yield of polymer being obtained in a shorter time.

EXAMPLE.

Ethylene is compressed under a pressure of 2000 atmospheres in a steel bomb. The vessel is maintained, in a constant temperature bath, for 4 hours at a temperature of 170° C. According to whether the pressure is released before or after cooling to room temperature, the product is respectively a " coalesced " rubber-like white solid or a white powder. The product softens at about 120° C. At 170° it goes over to the form of a substantially clear resin-like or rubber-like material which has plastic properties and can be moulded into various forms at or above the softening temperature. It is fairly soluble in hot benzene and sparingly soluble in cold benzene, hence thin transparent coatings may be deposited on objects from benzene solution, followed by warming until the deposit coalesces. The product has a molecular weight of around 3500 and in composition corre- §

[Price 1/-]

Radar

Means of detecting aircraft by using radio waves
Robert Alexander Watson-Watt for the National Physical Laboratory,
Teddington, Middlesex, England
Filed 17 September 1935 and published as **GB 593017**

Robert Watson-Watt was born in 1892. He was the Superintendent of the Radio Research Laboratory in Britain. The newly-created Committee for the Scientific Survey of Air Defence asked if radio could be used as a 'death-ray' to destroy enemy aircraft. The reply was given in a memo by Watson-Watt on 12 February 1935. The amount of power needed would make this impossible, but *detecting* aircraft would be possible by bouncing radio waves off them, and measuring the time of response, so that both direction and distance could be calculated.

The illustrated patent is from the first workable model. The Germans in fact anticipated the British work, with Rudolph Kühnold of the German Navy beginning work in 1933 and carrying out an experiment in Kiel harbour on 20 March 1934 with ships rather than aircraft. However, they only adopted pulse transmissions for calculating range in September 1935, something the British had realised was vital some months earlier. The British carried out a trial on 26 February 1935 where a Heyford bomber flew close to a transmitting station at Daventry at about 2,000 metres. It was identified at 13 km on a cathode-ray oscilloscope display in the back of a van. This experiment used only equipment readily to hand.

A purpose-built installation was next built on the coast at Orford Ness, in Suffolk. It was found that aircraft could be detected up to 64 km away. In December 1935 the government decided to build five stations to cover the approach to London up the Thames estuary. More were added and by early 1939 20 stations were operating from the Isle of Wight to Dundee. The Germans had a similar network along its North Sea coast. Although the stations were all large and obvious installations, neither country was aware of the work of the other.

The name 'radar' was coined by the Americans, for radio detection and ranging, in 1940. It was officially adopted by the British in 1943 to replace the earlier RDF, radio direction finding. Radar was vital in fighting off attacks from bombers during the Battle of Britain in 1940 as it meant that the relatively small Royal Air Force could direct its fighters to the most strategic places. Watson-Watt was knighted in 1942 but the patent was not published until 1947 on the grounds of secrecy. He died in 1999.

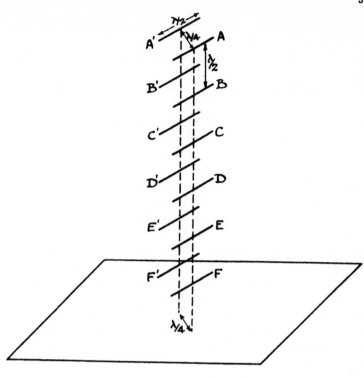

FIG. 6.

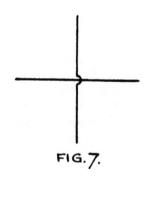

FIG. 7.

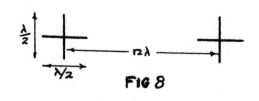

FIG 8

H.M.S.O.(Ty. P.)

Teflon®

Non-stick substance
Roy Plunkett, Wilmington, Delaware for Kinetic Chemicals (a subsidiary of Du Pont)
Filed 1 July 1939 and published as US 2230654 *and* GB 625348

Contrary to legend this material was not invented in the space programme. Roy Plunkett was working as a chemist for Du Pont and was asked to find a non-toxic refrigerant, the gas used in refrigerators to draw out the heat. He thought that he was mixing a quantity of tetrafluoroethylene, also known as Freon®. This should have resulted in a gas. When his assistant Jack Rebok opened the valve on the bottle the next morning no gas came out. They weighed the cylinder, and it definitely had something in it, as it weighed more than empty cylinders. The valve was open, as a wire went straight through it.

They had to saw the cylinder open to see what they had. A greasy white powder was inside. 'Gee whiz, it's gone wrong' said a perplexed Plunkett. Curious to know what he had, he gave the mysterious substance the usual tests: he rubbed it with his finger, tasted it, sniffed it, tried burning it, dropped acid on it. This was Teflon®, one of the first plastics. Like nylon, it is a polymer, with long chains of molecules. It was found to be inert: that is, nothing reacted with it so that heat, electricity and acids did not affect it. It was also very slippery. Du Pont could see many possibilities with such a substance.

Teflon® was kept secret until 1946 as it was considered so special. Its first practical use was when engineers working on the atomic bomb programme, the Manhattan Project, needed to protect gaskets from the very corrosive uranium hexafluoride used in making U-235. Teflon® did not come out in more than a few products until the late 1950s. It took time to work out methods to make it more cheaply and to make it into products. Muffin pans for bakeries were one of the first uses. Non-stick frying pans did not become available in America until 1960.

Those early pans had problems. They were easily scratched. So it was improved by bonding the Teflon® with the metal. Other products made from Teflon® include clothing, in electronics, and in medicine (the body doesn't reject it). Apparently Plunkett received no royalties. The name was registered as a trade mark in the United States in 1946 and in Britain in 1954.

UNITED STATES PATENT OFFICE

2,230,654

TETRAFLUOROETHYLENE POLYMERS

Roy J. Plunkett, Wilmington, Del., assignor to
Kinetic Chemicals, Inc., Wilmington, Del., a corporation of Delaware

No Drawing. Application July 1, 1939,
Serial No. 282,437

3 Claims. (Cl. 260—94)

A. This invention relates to new compositions of matter, being polymers of exceptional properties.

B. At the present time there are no totally satisfactory materials for handling certain corrosive agents, such as hydrofluoric acid, or for protecting workers against the fumes which arise from such reagents. Goggles having glass disks are attacked by the fumes, and shortly become unserviceable.

C. It is an object of the invention to provide a new composition of matter which is highly resistant to corrosive influences and to oxidation, and which can be molded and spun and put to a wide variety of uses where its peculiar properties would be advantageous.

D. The objects of the invention are accomplished by the compositions of matter which may be formed by the polymerization of tetrafluoroethylene. Other objects of the invention are accomplished by the process of polymerizing the fluoroethylene herein set forth.

E. I have discovered that tetrafluoroethylene will polymerize at ordinary temperatures when subjected to super-atmospheric pressure. I have also discovered that the rate of polymerization may be quickened by carrying out the polymerization under pressure in the presence of a catalyst. Furthermore, I have discovered that the polymerization of tetrafluoroethylene can be carried out advantageously in the presence of a solvent.

F. The following examples, which are summarized in the table, illustrate but do not limit the invention.

time the unpolymerized tetrafluoroethylene was removed, leaving a residue of 0.6 part of white solid polymer. The yield was 7.1% or a polymerization rate of 0.71% per day.

Example II

Tetrafluoroethylene (7.8 parts) was placed in a container under pressure at 20° C. The yield of polymer after 21 days was 0.05 part or 0.64%.

Example III

Tetrafluoroethylene (7.3 parts) was placed in a container with 0.1 part of zinc chloride, under pressure and maintained at a temperature of 20° C. The yield of polymer after 21 days was 0.1 part or 1.37%.

Example IV

Tetrafluoroethylene (5.4 parts) was placed in a container with 0.1 part of silver nitrate, under pressure at 25° C. After three days the container was completely filled with spongy white polymer. This material was partially polymerized tetrafluoroethylene, and had a very high vapor pressure. Yield of completely polymerized material was 0.05 part or 0.93%.

Example V

Tetrafluoroethylene (6.8 parts) was placed in a container with 0.1 part of silver nitrate under pressure at 25° C. The container was completely filled with partially and completely polymerized tetrafluoroethylene. Yield of completely polymerized product was 0.3 part or 4.4% in 10 days.

Table

Example	Parts C_2F_4	Time, days	Temp., °C.	Catalyst. and solvent for monomer	Yield, parts	Yield, percent
I	8.5	10	25	None	0.6	7.1
II	7.8	21	20	do	0.05	0.64
III	7.3	21	20	0.1 pt. ZnCl₂	0.1	1.37
IV	5.4	3	25	0.1 pt. AgNO₃	0.05	0.97
V	6.8	10	25	0.1 pt. AgNO₃	0.3	4.4
VI	7.0	21	25	0.1 pt. AgNO₃	2.3	33
VII	4.0		25	0.1 pt. AgNO₃, 2.5 methyl alcohol.	Jelly	
VIII	4.5	3	25	0.1 pt. AgNO₃, 2.2 methyl alcohol.	1.3	29
IX	7.4		25	0.1 pt. AgNO₃, 3.3 methyl alcohol.	2.0	27
X	8.8	21	25	0.1 benzoyl perox	0.05	0.57
XI	3.5		50	None		

Example I

Tetrafluoroethylene (8.5 parts) was placed in a steel cylinder under pressure and allowed to stand for 10 days at 25° C. At the end of this

Example VI

Tetrafluoroethylene (7 parts) and 0.1 part silver nitrate were placed in a container under

1940-1949

THE world was recast in the 1940s. The Yalta and Potsdam conferences divided Europe broadly in the geographic position that the victorious Allied countries had reached at the end of World War II in 1945. This partition, both real and symbolic, was to last for nearly 50 years.

The war's technology ensured there were millions of casualties. Whereas much of World War I had been a war of stalemate, World War II was a war of movement, of *blitzkrieg*, armour and airpower. This time millions of civilians too met their deaths by persecution or conquest or through horrific air bombing. Britain, France and Italy suffered less loss of life than in 1914-18 but the losses of Germany, Japan, Poland and—especially—the Soviet Union and China were enormous. The Allies' victory in Europe owed most to the fighting between the Soviet Union and Germany on the Eastern Front which saw the Soviet Union eventually flying its flag over Berlin. The scale of the largest individual battles there, at Stalingrad with more than 1 million casualties and at Kursk where 6,000 tanks and assault guns were engaged, dwarfed even the fierce battles like Caen in Normandy after D-Day, the Ardennes and Cassino in Italy. The Battle of Britain was an air triumph for Spitfire and Hurricane fighters and the naval successes of Midway and Leyte Gulf were proof of the United States' sea and air power. In the Far East the war was ended by the USA's atom bombs falling on Japan.

Six million European Jews were murdered in Nazi-held territories and many other groups and races killed with them. This became central to the belief of so many that true evil was the ultimate enemy which had to be defeated in World War II. And unlike World War I the leaders of the chief powers with their different philosophies and ideologies—Churchill, Roosevelt, Stalin for the Allies; Hitler, Mussolini, Tojo for the Axis—appeared to take a much greater role in the centre of the historical stage. These figures became more identified in the popular mind, than the Great War leaders had been, with the justice or injustice of the causes of the war of 1939-45 and what might follow it.

As the 'great arsenal of democracy' the resources of the USA were crucial to land, sea and air victory in Europe and the Pacific. The atom bomb introduced the nuclear age and 1945 began an area of confrontation, but also global balance, between the superpowers of the USA and the Soviet Union. Mutual suspicion in this cold war resulted in the superpowers backing rival states around the world. The desire to band together to meet a possible threat from communist countries encouraged Western states to cooperate with each other as illustrated by the remarkable Berlin airlift in 1948 which overcame a Soviet transport blockade of the divided German city.

Cooperation also brought new structures. Just as World War I had created the League of Nations so World War II created the United Nations. New power blocs such as the NATO alliance were established as well as organisations to bring further economic, social and cultural benefits like the Council of Europe and the Organisation for European Economic Cooperation. Nor was the vindictive folly of the treaty of 1919 repeated. The Marshall Plan for economic recovery ensured a massive flow of funding from the USA to Europe and both Germany and Japan began to rebuild their economies and industries with aid from the war's victors. The Bretton Woods conference agreed the need to form a future World Bank and International Monetary Fund.

The breaking up of the world's colonial systems was hastened in Asia by the independence of India, Pakistan, Indonesia, and Burma while the communist cause triumphed after civil war in China. The state of Israel was born, and baptised in war with its Arab neighbour states. This created the tensions in the Middle East which endure. The decade which presided over the first atomic explosion and the start of the nuclear arms race saw great advances in science and technology like Fermi's nuclear reactor and the work of his colleagues at Los Alamos, von Braun's rockets, the first jet airfighter (Messerschmitt Me262), the first jet airliner (De Havilland Comet) and an aircraft breaking through the sound barrier.

The onset of computing was given added impetus by wartime research on military communication codes: British unscrambling of German secret codes had helped the Soviet Union prepare for the battle of Kursk. Magnetic tape was invented and assembler programming language compiled for computers. Hillier produced his electron microscope and there were early experiments with colour television. The LSD drug was discovered, quinine synthesised, medicinal properties of cortisone identified and the citric acid cycle discovered by Krebs. In the midst of the many tragedies which clouded the 1940s the long-playing sound record and the scuba aqualung made their first appearance to delight, then and later, so many people of different races and nations.

The ballpoint pen

Pen with a rotatable ball as a nib
Laszlo Josef Biro, Buenos Aires, Argentina
Filed 10 June 1943 and published as **GB 564172**

The principle of the ballpoint pen dates back to John Loud of Massachusetts with his US 392046 in 1888, and includes Biro's own GB 498997, but this is the first practical ballpoint pen. Laszlo Biro was a Hungarian journalist (and also a sculptor and hypnotist). On a visit to a newspaper printing press in 1938 he noticed that the ink used dried quickly and left no smudges. The ink was thicker than that used in fountain pens (which easily blotted) and wouldn't flow from the nib of fountain pens. He began to work on a way to use such ink in pens. In the meantime he and his brother George, a chemist, emigrated to Argentina in 1940.

The idea of pens with freely moving balls as a nib was known but it had not been possible to devise a suitable feed of ink. After years of experimentation they came up with a pen where the ball rotates when in use and has a supply of ink from a capillary constriction that flows down grooves in the ball only when pressure is applied, as when writing is carried out. This design meant that it would not leak. It also did not need frequent refills.

Soon after they filed the patents a British government official, Henry Martin, visited Buenos Aires. He heard about the pen and bought the British licensing rights. Air crews would find them useful for navigational work as such pens wouldn't leak at high altitudes, unlike fountain pens. Manufacture was begun in a disused aircraft hanger near Reading in 1944.

Meanwhile, the Eterpen Company began manufacture in Argentina. They were sold for the equivalent of £27. In June 1945 Milton Reynolds, a Chicago businessman, was in Buenos Aires. He noticed the product in a shop and purchased some samples. He employed William Huenergardt to come up with an imitation, Reynolds' Rocket. They filed for a patent in September 1945 but dropped it and refiled for the improved US 2462453 three months later. This pen worked somewhat differently. The product was launched in October 1945 for $12.50 and 10,000 were sold in a day.

Biro was about to launch his own product through Eversharp. According to legend Biro had failed to take out American patents although he did have US 2390636 (filed in 1943) and US 2400679, which perhaps were earlier forms. Court action failed and sales took off for both products. However, both had problems, especially the Reynolds product, with the ink skipping. As both offered to take back defective products losses soon mounted. In 1951 Reynolds' company folded while Eversharp struggled on for some years. In 1953 Baron Bic introduced the 'throwaway' ballpoint in European markets. In America Parker's Jotter pen came out in 1954. Both quickly sold many millions.

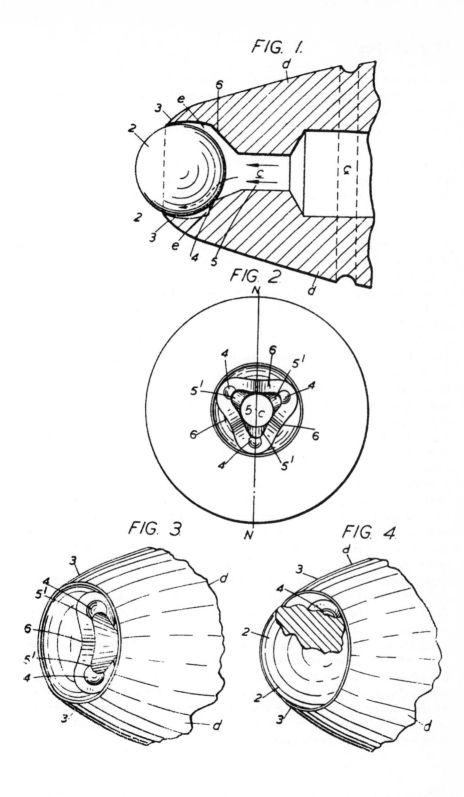

FIG. 1.

FIG. 2.

FIG. 3

FIG. 4.

Bar codes

Data on products identifying them for reading by scanners
Norman Woodland, Ventnor, New Jersey and Bernard Silver, Philadelphia,
Pennsylvania
Filed 20 October 1949 and published as US 2612994

Bernard Silver was a graduate student at Drexel Institute of Technology (in Philadelphia). One day he overheard the president of a local food store chain asking one of the deans if the Institute could come up with a way of capturing product information automatically at the checkout (he was turned down). Silver was intrigued and told his fellow graduate student, Norman Woodland, about it. Woodland was excited and took the lead in research. First he tried using patterns of ink that would glow under ultraviolet light, but the ink was unstable and it was an expensive way of labelling. He sold his stockmarket shares, left graduate school and went to his grandfather's apartment in Florida to think about it for a few months.

He decided to use an idea from films to scan lines taken from the Morse code— dot, dash where each was extended to form thick or thin lines. With four white lines on a dark background there would be scope for seven combinations: with ten, 1,023. He later adapted the idea to a bull's eye, as shown in the drawing, as there would be no need to 'line up' the article, which could be viewed from any angle. The patent states that it could be used in 'super-stores' but does not limit it to that use. A bar code was no use without a cheap and reliable scanner. Woodland got a job at IBM and he and Silver worked on a scanner in their spare time. They built a scanner the size of a desk, with a 500 watt bulb and an oilcloth covering it to keep out the light. When they tried it, the paper samples smouldered in the intense heat, but it did work.

IBM twice offered to buy the patent, but they were turned down as they did not offer enough. In 1962 Philco bought the rights, which they later sold to RCA. Silver died the following year without seeing his idea in practice. With rapid improvements in computer power some experimental implementations were tried out, and much interest was aroused. IBM was envious of RCA's work on bullseye bar codes, though the variant was later dropped for the modern parallel lines type because the ink could smear between the different fields of data. Then someone mentioned that Woodland was still on the staff. He helped design the Universal Product Code in 1973, which standardised usage of digits and gave a cue to tell the scanner which way to read them.

Although scanners using lasers to read the codes were still relatively expensive large stores and chains could see the benefits in cutting down time spent at the checkout and in getting better control of inventory. Mistakes were also much less likely to happen. Even before bar codes were used in stores the Universal Grocery Products Identification Code was devised in 1970 giving a unique number to each product. On 26 June 1974 in Troy, Ohio the first product was scanned at a supermarket—a packet of chewing gum.

FIG. 1

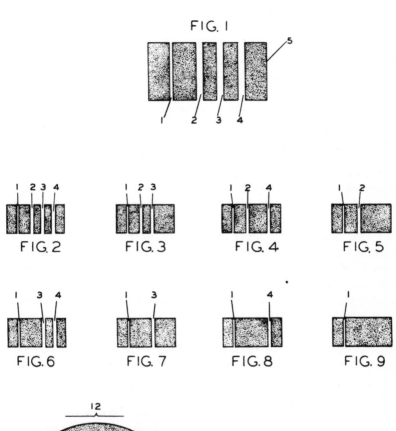

FIG. 2   FIG. 3   FIG. 4   FIG. 5

FIG. 6   FIG. 7   FIG. 8   FIG. 9

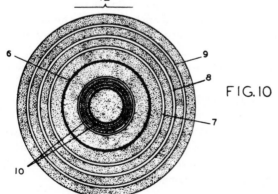

FIG. 10

NOTE: LINES 6, 7, 8, AND 9 ARE LESS
REFLECTIVE THAN LINES 10.

INVENTORS:
NORMAN J. WOODLAND
BERNARD SILVER
BY THEIR ATTORNEYS
Howson &
Howson

The computer

Electronic apparatus for making calculating or controlling operations
John Eckert, Jr. and John Mauchly, both of Philadelphia, Pennsylvania, for
Sperry Rand Corporation
Filed 26 June 1947 and published as US 3120606 *and* GB 709407

Giving the inventor of the computer is impossible as it depends on what you mean
by a computer. The term comes from that for mathematicians who were skilled at
doing calculations. The history is so complex that this can only be a sketchy summary, but it is interesting that most early work was directed towards calculations,
which are easier to programme than other aspects of computers. There were early
pioneers such as Hermann Hollerith, the US census clerk whose 1889 and 1890
patents were used to sort automatically cards bearing census data in the 1890 census. In 1931 the German engineer Konrad Zuse made his Z1 computer which performed calculations using binary code (zeroes and ones, represented by the presence
or absence of an electric charge). This is how modern computers work. He did use
electromechanical relays rather than electronics, which he proposed in 1940. His developing ideas and machines were not actively supported by the German government during World War II, which hampered his progress.

Meanwhile, in 1939 in Iowa John Vincent Atanasoff, assisted by Clifford Berry,
devised the ABC computer. This used vacuum tubes as well as binary codes to do
calculations. This was not patented, and he showed what he was doing in 1941 to
John Mauchly and John Eckert from the University of Pennsylvania. From 1943 to
1946 they built the ENIAC machine, which weighed 30.5 tonnes and had 18,000
vacuum tubes. On average it could perform calculations for 7 minutes before a tube
broke. Its memory was 16K and every time they needed to do a fresh programme
they had to rewire the entire machine. To make it more impressive at a press conference ping pong balls were cut in half and placed above lights on the machine so
that they glowed.

Harvard's Mark 1 machine was the first to have stored instructions on paper tape.
Occasionally it had to be stopped for repairs, and once a moth was found in one of
the relays. After that 'debugging' was mentioned when repairs were needed which
is thought to be the origin of the word. While all this was going on, the British had
built the Colossus machine in 1943. Thomas Flower of the Post Office built it so
that German intelligence messages could be decoded. It was kept secret until 1976.
Some have suggested that Colossus should really be considered the first computer.
In 1950 Remington Rand launched the Univac 1 computer, which was the first commercial model. Forty-six were built but it never made money. They were used in
the 1950 census. The illustrated patent was published as late as 1964 although the
corresponding British patent was published in 1954. This gives a different applicant, the Eckert-Mauchly Computer Corporation.

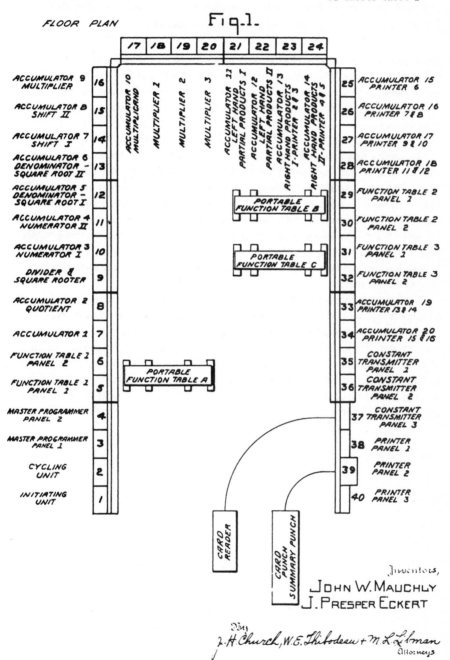

FLOOR PLAN　　　Fig.1.

The hologram

Three-dimensional imaging
Dennis Gabor for the British Thomson-Houston Company, London, England
Filed 17 December 1947 and published as **GB 685286**

Dennis Gabor was born in Budapest in 1900. An electrical engineer, he went to live in Germany in 1927 and worked for Siemens & Halske but fled to England after the Nazis came to power. His interest was in 'seeing' individual atoms. The electron microscopes of the time could not resolve sharp images. He was sitting on a bench at a tennis club in 1947 when he thought of an electron image that could be corrected by optical images—a 'hologram', Greek for 'completely written'.

His model had in fact limited use as a powerful and focused light source was needed, which was later provided by lasers, normally with the continuous-wave laser. The way it works is as follows. In a darkened room a beam of coherent laser light is projected onto a picture which is scattered back to form an image on a photographic plate. Simultaneously part of the laser beam is reflected by a mirror onto the same plate. The two images 'interfere' with each other there, forming 'interference fringes' in a pattern of alternating light and dark. The light is where the two images both reflect light back and reinforce each other, while the dark is where the images do not match.

The image is reconstructed for the hologram by reversing the procedure so that a three-dimensional picture is achieved which varies as you move your eyes. Holograms with their colourful, glossy images are used a great deal in advertising and entertainment. Emmett Leith and Juris Upatnieks with their US 3506327, published in 1970, applied laser beams to holograms. Gabor joined the faculty of the Imperial College of Science and Technology in London in 1949, teaching applied electron physics. He received over 100 patents. He was awarded the Nobel Prize for Physics in 1971 and died in 1979.

685,286 PROVISIONAL SPECIFICATION
I SHEET   This drawing is a reproduction of
the Original on a reduced scale.

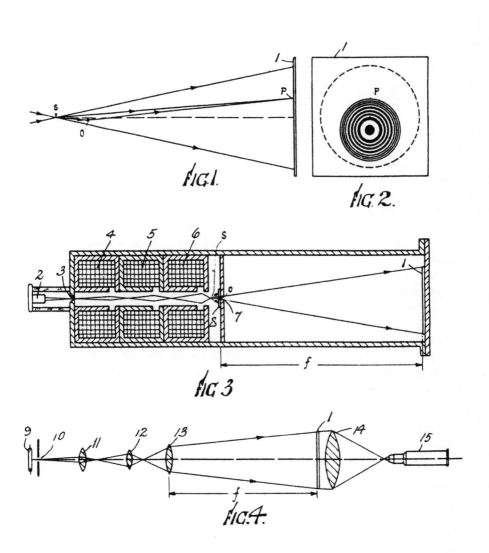

FIG.1.

FIG. 2.

FIG 3

FIG.4.

Instant photography

Camera which rapidly develops pictures
Edwin Herbert Land, Cambridge, Massachusetts, for Polaroid Corporation
Filed 3 February 1948 and published as US 2543180

Edwin Land was still a physics student at Harvard when he succeeded in producing the first modern filters to polarise light. This has applications in sunglasses, headlights and other requirements in reducing glare. He formed the Polaroid Corporation in 1937 which had an active research programme in optical devices. One day his three-year old daughter asked why she could not see a photograph right away that he had just taken of her. Land went away and created both a camera, illustrated here, and the special film required (filed 11 December 1948 and published as US 2543181).

Marketed as the Polaroid Land Instant Camera, at the premium price of $89.50, the new product was launched at a Boston department store in 1948. It was a great success and 1 million had been sold by 1956. The new camera was able to produce a dry black and white photograph in about a minute. The film had to be peeled off a backing. It worked by reproducing the image recorded by the lens directly onto a surface which acted as both the film and as the finished product. Many improvements were made over the years. In 1965 a big success was achieved with the Swinger, which produced a small black and white photograph and cost only $19.95.

The film for these cameras was produced by the Kodak Corporation, but they wanted to enter the market themselves. In 1969 they stopped supplying film. Polaroid, aware of Kodak's interest in the market, came out with the SX-70 camera in 1972 which for the first time simply produced a photograph without any peeling away of backing paper.

In 1976 Kodak introduced an instant camera. They were charged by Polaroid with infringing ten patents. By this time most of the company's business was in instant pictures and it was crucial to them to keep their market share. Each camera was designed so that only the same company's film would be accepted. The case was decided in 1985 with seven patents declared infringed. An injunction stopped Kodak from making instant cameras. It was not until 1991 that the case was finally settled. Kodak paid $925 million in damages, a record sum for the time. Land himself also died in 1991, having received over 500 patents.

Feb. 27, 1951 E. H. LAND 2,543,180

PHOTOGRAPHIC APPARATUS

Filed Feb. 3, 1948

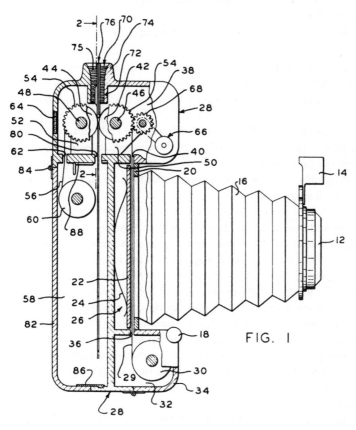

FIG. 1

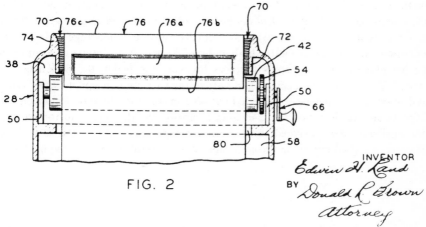

FIG. 2

INVENTOR
Edwin H. Land
BY Donald L. Brown
Attorney

The microwave oven

Oven using microwave energy to cook food
Percy Spencer, West Newton for Raytheon Manufacturing Company, Newton, both Massachusetts
Filed 8 October 1945 and published as US 2495429

At the beginning of World War II Raytheon was a small company working in military electronics. When the British invented the magnetron, which is at the heart of radar, they found it difficult to make improvements. Raytheon managed to get invited to confidential talks with British experts. Percy Spencer told these experts that their methods of making the tubes were 'awkward and impractical'. He persuaded them to let him take one home and over the weekend came up with suggestions that simplified the manufacture and incidentally improved the performance. The company was initially awarded a small contract and by the end of the war was producing 80% of all magnetrons.

One day in 1945 Spencer walked past a switched-on magnetron and realised that a chocolate and peanut bar in his pocket had been turned into a gooey mess. He felt no heat, and realised that it was the microwaves from the magnetron that had cooked it. He sent out for a bag of popcorn and put it in front of the magnetron, and popcorn exploded all over the laboratory. The next day he put a raw egg in a kettle with one side cut out and placed it in front of the magnetron. An impatient colleague looked over at the hole at the moment that the egg exploded and literally got egg in the face (the inside had cooked faster than the shell). Normal cooking occurs by working in from the outside (convection). Microwave ovens work by agitating the water molecules in the food, and such movement is a normal sign of heat. Because of the way it works it does mean that more food takes longer to heat up (there is more material for the oven to work on) and the inside of the food is cooked at the same time as the outside.

The patent envisages the food passing the oven as shown by the conveyor system on the right of the drawing. The speed of the movement would determine how long the cooking lasted. Raytheon had not been involved in the consumer market but decided to see if it could market what was initially called its 'Radarange'. The original microwave ovens were 2 metres high and cost $3,000. A prototype was put in the kitchen of one of the company's directors: his cook, horrified by the black magic, resigned. With such a high cost it sold only to catering units, hospitals, army camps and so on where cooking was carried out on a major scale. In 1965 Raytheon took over Amana Refrigeration Unit, an Iowa company, and they began in 1967 to market smaller models designed to fit on a worktop, when sales began to pick up. Spencer, a self-educated man from Maine, was responsible for over 200 patents.

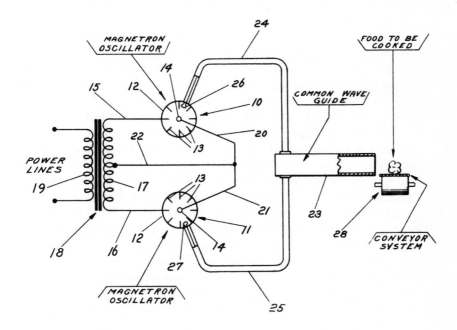

INVENTOR.

PERCY L. SPENCER,

BY Elmer J. Gorm
ATTY.

Silly Putty® material

Moulding material that bounces
Rob Roy McGregor, Verona and Earl Warrick, Pittsburgh, Pennsylvania for
Corning Glass Works, Corning, New York
Filed 30 March 1943 and published as US 2431878

Silly Putty® material is one of those inventions that have had two claimants to its invention, although it is clear that one man was responsible for its success. McGregor and Warrick were scientists working for Corning Glass. Rubber was in short supply in World War II, and they were looking for a way to make synthetic rubber with silicone, a derivative of silicon (that is, sand). In an experiment, boric acid was added to a fluid called Corning 200 fluid, which looks like petroleum. It was heated in an oven overnight, and a strange substance was found the next day. They quickly realised that they could have fun moulding, stretching and bouncing it round the laboratory, but it was useless as a rubber. If they put it in water the boron came out, and the original oil was left. They kept some as a curiosity and filed a patent anyway for this method of 'treating dimethyl silicone polymer with boric oxide'.

A little later, James Wright of General Electric was also trying to figure out how to make synthetic rubber from silicone. The company claims that he too mixed up boric acid with silicone and got something unusual. When someone asked what good it was a chemist on the team pointed out that you could drop it on the floor and say 'Golly, look at it bounce!' General Electric sent out samples worldwide asking if anyone could think of a use for it, but no one could. They too filed for a patent, published as US 2541851. The putty occasionally surfaced at parties involving scientists, when people played around with it. Ruth Fallgatter, owner of a toy store in New Haven, Connecticut, hired Peter Hodgson, an advertising copywriter, to produce a catalogue which included the putty. A description of it was put on the same page as a spaghetti-making machine and a roller. The price was $2, and it was available in a small, clear case.

The product outsold everything in the catalogue except for a crayon set. Fallgatter lost interest, but Hodgson decided that it was an interesting product. He borrowed $147 and bought all he could of the putty and put 28 gramme portions into clear plastic eggs priced at $1. He introduced it at the 1950 International Toy Fair as Silly Putty® (which he registered as a trade mark in 1952) but there was no interest from the experts. He then tried Neiman-Marcus, the department store, and Doubleday bookshops. The eggs were shipped to them in old egg crates. In 1950 a writer for the *New Yorker* magazine saw the eggs in a Doubleday store. He wrote about them in its 'Talk of the town' column and within three days Hodgson received orders for over 750,000 eggs. He also waged one of the first children's television advertising campaigns. Over 230 million have been sold since, and at Hodgson's death in 1976 he was worth $140 million. The putty even went into space for the twin purposes of alleviating astronauts' boredom and to fasten things down.

UNITED STATES PATENT OFFICE

2,431,878

TREATING DIMETHYL SILICONE POLYMER WITH BORIC OXIDE

Rob Roy McGregor, Verona, and Earl Leathen Warrick, Pittsburgh, Pa., assignors to Corning Glass Works, Corning, N. Y., a corporation of New York

No Drawing. Application March 30, 1943, Serial No. 481,144

6 Claims. (Cl. 260—46.5)

1

This invention relates to new compositions of matter, their preparation and uses, and, more particularly, to organo-silicon polymers and methods of preparing them.

The present invention is concerned with a method of further polymerizing organo-silicon oxide polymers and with the products derived therefrom. Organo-silicon oxide polymers are compounds which contain organic radicals attached to silicon through a carbon atom and whose silicon atoms are joined to other silicon atoms by oxygen atoms, thus Si—O—Si. They may be prepared by the hydrolysis of hydrolyzable organo-silicanes and condensation of the hydrolysis products. Furthermore, hydrolysis of a mixture of different hydrolyzable organo-silicanes and co-condensation of the hydrolysis products produces organo-silicon oxide copolymers which are within the scope of our invention. In the latter case, a hydrolyzable silicane which contains no organic radicals attached to silicon through a carbon atom, such as silicon tetrachloride or ethyl orthosilicate, may be included along with the hydrolyzable organo-silicanes. By hydrolyzable organo-silicanes we mean derivatives of SiH_4 which contain readily hydrolyzable radicals such as hydrogen, halogens, amino groups, alkoxy, aroxy and acyloxy radicals, etc., the remaining valences of the silicon atoms being satisfied by organic radicals that are joined to the silicon atoms through carbon atoms such as alkyl, substituted alkyl, aryl, substituted aryl radicals, etc.

Hydrolysis of the above silicanes or mixtures thereof is generally accompanied by condensation to a greater or less degree depending upon the conditions of hydrolysis and the particular silicanes involved. As a result of the hydrolysis and concurrent condensation, organo-silicon oxide polymers or organo-siloxanes (as they are now commonly called) are produced which are partially or completely condensed and which have on the average up to and including three organic radicals attached to each silicon atom. The polymers so obtained vary in character, some being oily liquids, others being crystalline solids or gels. They also vary in the ease with which they may be further polymerized by heat since they differ in the number of active functional groups retained as a result of incomplete hydrolysis and condensation. Those which are only partially condensed may be converted to higher polymers and even to solids by heat alone or even by standing at room temperature by virtue of the completion of condensation. On

2

the other hand, those organo-siloxanes which approach complete condensation are extremely resistant to further polymerization by heat alone. These substantially completely condensed polymers are not limited to those which are of high molecular weight but may be polymers of low molecular weight as well. For example, the condensed hydrolysis products of the di-organo-silicanes are essentially completely condensed even in the low polymeric stages and exist generally in the trimeric form with polymers as high as the hexamer being reported in only rare instances. Since the higher polymers of these organo-silicon oxide compounds, and particularly the higher polymers of the substantially completely condensed compounds, have been found to possess properties which adapt them to many industrial applications as will be described below, it is highly desirable to provide a method of further polymerizing the organo-silicon oxide polymers to higher polymeric compositions, that is, to increase their average molecular weight.

The primary object of this invention is to provide a method of polymerizing the hydrolysis products of hydrolyzable organo-silicanes or mixtures thereof.

Another object of our invention is to provide a method of further polymerizing an organo-siloxane having on the average less than three organic radicals attached through carbon atoms to each silicon atom.

Another object of our invention is to provide a method of further polymerizing a substantially completely condensed liquid hydrolysis product of a silicane of the type R_2SiX_2, where each R is an organic radical which is joined to the silicon atom through a carbon atom and each X is a hydrolyzable atom or group.

Still another object of the present invention is to provide a method of polymerizing a substantially completely condensed liquid hydrolysis product of a mixture of silicanes comprising essentially a di-organo-substituted silicane to a polymeric composition which is substantially free of polymers having less than seven silicon atoms per molecule.

Another object of our invention is to prepare organo-silicones of high average molecular weight.

In accordance with our invention, we have provided a method of polymerizing organo-siloxanes having on the average less than three organic radicals attached to each silicon atom which comprises maintaining boric oxide in intimate and continuous contact with said polymers. If de-

Slinky®

Toy
Richard James, Upper Darby, Pennsylvania for James Industries
Filed 1 November 1945 and published as GB 630702 *and* US 2415012

This is the familiar toy which consists of coils that move downstairs, along the floor, or from hand to hand. Richard James was a mechanical engineer working for the US Navy. While on a ship undergoing trials, a lurch caused a torsion spring to fall accidentally from a table to the floor. Its springy movement made him think. When he saw his wife Betty that night he showed her the spring and said 'I think there might be a toy in this'. Two years of experimentation followed to achieve the right tension, wire width and diameter. The result was a steel coil with a pleasant feeling when handheld, with an ability to creep like a caterpillar down inclined planes or stairs, and an interesting action when propelled along the floor. Betty came up with the name of Slinky®, from slithering.

James managed to persuade Gimbels, the department store, to give him some space at the end of a counter. He would demonstrate the toy and hope to sell some of his stock of 400. It was a miserable November night, and Betty and a friend were on hand to buy a couple to encourage sales. They never had the chance, as crowds gathered around and the entire stock went in an hour and a half. A company, James Industries, was set up to make the product. A machine was devised which coiled 24 metres in 10 seconds. The price was $1 in 1945, which had increased to $2 by 1994. Over 250 million have been made, with some variations, including brightly coloured plastic models. The only substantial change in the design is that the end wires are now joined together to prevent loose wires damaging, for example, an eye. The trade mark was registered in the United States in 1947 and in Britain in 1946.

Richard James became interested in a cult and gave substantial amounts of the profits to them. He later went to Bolivia to join the cult. Betty then took over the company, reorganised it and continued to run it as an independent company, despite offers from many toy manufacturers.

Besides the obvious fun possibilities, the toy has been used by science teachers to demonstrate the properties of waves. NASA has used them to carry out zero-gravity physics experiments in the space shuttle. And in Vietnam, American troops used them as mobile radio antennae.

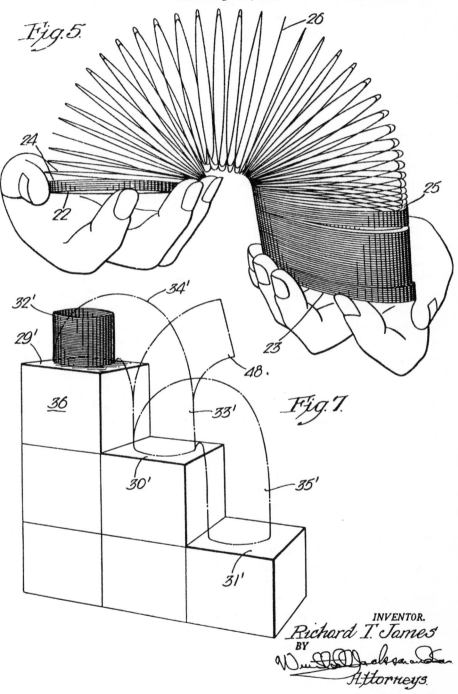

Fig.5.

Fig.7.

INVENTOR.

Richard T. James

BY

Attorneys.

The transistor

Transistor using a semiconductor to control or amplify small electric currents
John Bardeen, Summit, and Walter Brattain, Morristown, both New Jersey,
for Bell Telephone, New York
Filed 17 June 1948 and published as US 2524035 *and* GB 694021

The old radio sets and many other electronic devices were large as they contained vacuum tubes (which control electric currents); these were both bulky and required heating to work, so that a heat source was needed. A smaller solution, which did not involve heating, was necessary for small electronic products. Before World War II, Bell Telephone began to study solid-state physics with an emphasis on semi-conductors. Semi-conductors were known and to some extent already used—they are materials which conduct weak electric currents. William Shockley predicted that an electric charge applied from outside could move the electrons in a semi-conductor so that it acted as an amplifier. The war interrupted the research, but during the war other discoveries were made, such as Purdue University finding that germanium was a useful material.

As soon as the war was over Shockley organised his team. He found that his theory did not work: the electric current could not penetrate the semi-conductor. John Bardeen, a theoretical physicist, had a theory about the nature of the surface of the semi-conductor which would account for this. A test with inferior films showed that the current would go through if it were transmitted through an electrolyte in contact with the surface. However, this did not work with the preferred germanium as the effect was the opposite to that predicted. After experiments this was modified to a point-contact transistor, where current flowing to one contact is controlled by current flowing from a second contact. All this was done with relatively simple equipment. This first transistor was noisy, could not control high amounts of power and had a limited applicability. Shockley thought up the idea of the junction transistor, which overcame most of the problems. Other improvements included replacing germanium with silicon.

There was a dispute over who ought to be credited as the inventor. Shockley believed that as he had the original idea he ought to get sole credit, and called Bardeen and Brittain in separately to say so. Bardeen stormed out, while Brittain said 'Oh hell, Shockley, there's enough glory in this for everyone!'. Bell planned for a patent in Shockley's name but discovered that the prolific Julius Lilienfeld had patented something virtually identical to Shockley's idea. So the original point contact transistor patent was based on Bardeen and Brittain's work, while Shockley received US 2569347 for his improved junction transistor. All three received the Nobel Prize for Physics in 1956. Bardeen also shared it in 1972 for superconducting. By the late 1960s transistors were beginning to be superseded by integrated circuits, which involve many transistors and other components on a single wafer.

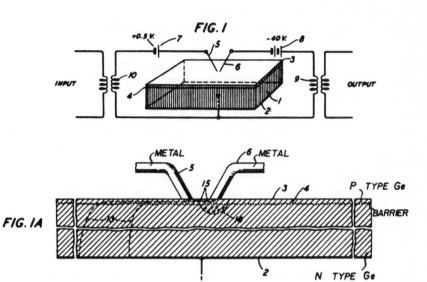

Oct. 3, 1950 J. BARDEEN ET AL 2,524,035

THREE-ELECTRODE CIRCUIT ELEMENT UTILIZING SEMICONDUCTIVE MATERIALS

Filed June 17, 1948 3 Sheets-Sheet 1

FIG. 1

FIG. 1A

FIG. 2

FIG. 10

FIG. 11

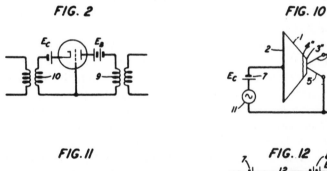

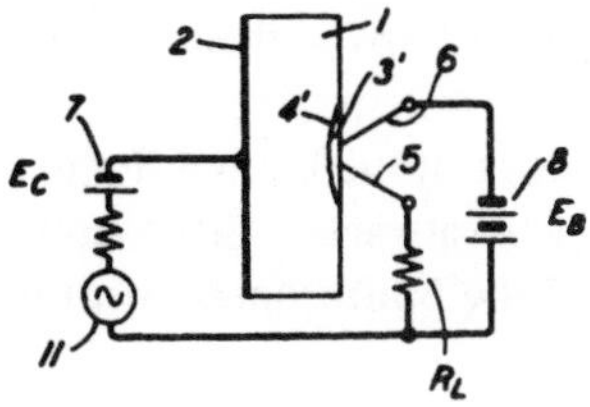

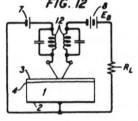

FIG. 12

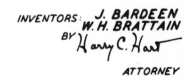

INVENTORS: J. BARDEEN
W. H. BRATTAIN
BY Harry C. Hart
ATTORNEY

Tupperware®

Sealable plastic containers
Earl Silas Tupper, Upton, Massachusetts
Filed 2 June 1947 and published as **US 2487400** *and* **GB 662219**

Earl Tupper was an engineer working in a Du Pont chemical plant. He was convinced that plastic was the material of the future, and left to form the Tupper Plastics Company in 1938. He asked Du Pont for some material with which he could experiment. They gave him polyethylene slag, a leftover from oil refining. It was a black, hard material. Tupper refined it and managed to make a semi-transparent substance that could be moulded by machinery without cracking or splitting. He thought of products that could be made from it. Shoe heels were one possibility; so were cups and bowls, and containers. Consumers were not impressed by his products as plastic had a poor image, even though his line was of high-quality, strong and simple products. Glass or pottery was used at the time, and they were neither airtight nor shatterproof.

Tupper realised that food could be stored without drying out in his containers if he could make an airtight seal. The patent is for this seal: the rim of the containers flares outwards slightly, and the lid, which is of the same dimensions as the container, snaps securely once the air inside is 'burped out'. Only plastic offered the right qualities of recovery for repeated closures, so that it would not break under the repetitive strain. The result was something that would not spill or break, even if dropped, and would keep food fresh for a long time.

Sales in the shops were still not very good, often because people could not work the required 'burp'. By chance, a woman called Brownie Wise was given one by a friend. It took her three days to work out how to 'burp' it properly. She dropped it on the way to the refrigerator, and was pleased that this novel product did not burst. Wise was a distributor for another company and asked if she could demonstrate the product at home parties, where people were invited to a house to try the product. Over a 2 hour session she would demonstrate the 'burp' and the guests would, hopefully, clamour to buy at the end. This was the beginning of the famous 'Tupperware home parties'. She began to make sales of up to $1,500 a week this way. Earl Tupper took notice and stopped selling through the shops and hired numerous demonstrators. The product was on its way.

The trade mark Tupperware® was registered in 1951 in the United States and in Britain as late as 1965. The containers are considered so attractive that examples are in the Museum of Modern Art in New York and in the Smithsonian Institution in Washington, D.C.

Nov. 8, 1949

E. S. TUPPER

OPEN MOUTH CONTAINER AND NONSNAP
TYPE OF CLOSURE THEREFOR

2,487,400

Filed June 2, 1947

2 Sheets—Sheet 1

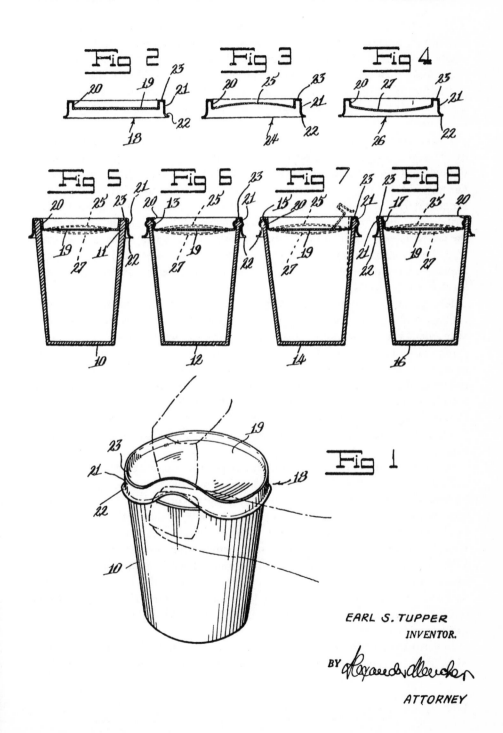

EARL S. TUPPER
INVENTOR.

BY

ATTORNEY

1950-1959

THE second half of the century was to experience great progress, especially in electronics, nuclear and space technology, medical and genetic engineering. It would also spur many more inventions that granted greater leisure to people or could make their leisure time more enjoyable. And it would give rise to movements, like the campaign for nuclear disarmament, whose adherents were alarmed by the speed and reach of technological developments, especially if they were in rogue hands.

Reconstruction in the West had been aided by the Marshall Plan and in the East by the Soviet Union's 5-year planning linked to its alignment with the economies of satellite states in Eastern Europe. The Soviet Union did not hesitate to maintain this link forcibly, as in Hungary in 1956. The Colombo plan too started to funnel aid from the USA to countries in South Asia. The end of food rationing in Britain was one symbol of the far greater prosperity that began in the 1950s for the industrialised world and for countries with more efficient systems of agriculture. The creation of a European Coal and Steel Community and the European Economic Community (the Common Market) by the treaties of Paris and Rome accelerated this trend.

Khrushchev's government in the Soviet Union sought to introduce some liberalisation of Soviet society and a gradual process of de-Stalinisation. The Cold War among the superpowers remained icy, however, and suspicion of increasing communist influence in the USA led to McCarthy's crusade against possible communist supporters there. The full tide of this anti-communism in the USA began to ebb in the late 1950s; at the same time the civil rights movement for racial equality gathered force. The Warsaw Pact between the Soviet Union and its dependent states was convened as a counter to the NATO alliance which West Germany had joined. But the summit conferences of world leaders now also began.

Harry Truman's Doctrine to 'support free peoples who are resisting attempted subjugation' ensured that North Korea's invasion of the South would be resisted by troops of the United Nations. The inconclusive Korean War became the first bitter battleground fought over by superpower blocs, including the new communist regime in China; major combat with jet aircraft first occurred in Korea. Europe's footholds in Asia reduced still further when France withdrew from Vietnam after conceding defeat to the Vietminh at Dien Bien Phu. There was resistance to British rule in Cyprus in support of union with Greece, and the attempt in 1956 by Britain and France to force Egypt to reopen the Suez Canal and to return its revenues was to bring the curtain down on the last act of Britain's role as a semi-imperial world power.

Independence or the struggle for it spread across Africa: in Ghana, in violent Mau Mau uprising in Kenya and in Algeria's rebellion against France. The

Algerian fighting brought France's 1940s statesman, Charles de Gaulle, to power once more and this was also to lead to growing French influence in Europe. Other leaders came to world notice, or began to be political icons for many years to come, like Nasser in Egypt, Castro and Che Guevara in Cuba and Ho Chi-Minh in Vietnam. But Juan Perón, whose regime together with his wife Eva had rekindled international interest in South American affairs, was overthrown from Argentina's presidency: only to return to it 20 years later.

This period saw a first hydrogen bomb exploded on Bikini Atoll, the first computer to be produced commercially, the Sputnik and Explorer satellites launched into space (and the NASA space agency established) and the structure of DNA discovered with its profound implications for so much genetic and medical research since. Energy could now be provided via a nuclear power station. An Wang invented magnetic core computer memory and the Fortran computer programming language was devised. Bell Telephone developed an early visual telephone, Boeing introduced its 707 'big jet' airliner, Rolls-Royce produced the first vertical take-off aeroplane prototype ('the flying bedstead') and Pilkington the float glass process for making plate glass. Greater wealth and leisure time in developed countries enabled people to take advantage of the first appearance in the 1950s of a credit card, transistor radio and video recorder and to visit the first 'Disneyland' theme park.

The conquest of nature and the elements was yet further fulfilled by the opening of the St Lawrence Seaway and the journey of the nuclear submarine *Nautilus* beneath the North Pole's ice cap. The successful ascent to the summit of Mount Everest in 1953 was a feat which many thought surpassed even Amundsen's journey to the South Pole of 40 years earlier.

Float glass

Using a bath of molten tin to make plate glass cheaply
Lionel Alistair Pilkington, Rainhill and Kenneth Bickerstaff, St Helens for
Pilkington Brothers, Liverpool, all in Lancashire, England
Filed 10 December 1953 and published as **GB 769692** *and* **US 2911759**

Traditionally there were two ways of making plate glass. 'Window glass' involved forming a sheet by stretching molten glass either by blowing or by pulling. Distortion tended to occur, but the product was cheap to make. 'Plate glass' was made in various ways but all involved casting a plate of glass, grinding it flat, and then polishing it to make it transparent. This was expensive and there was a loss of glass when it was polished.

Float glass involves a ribbon of glass moving on rollers from a furnace at 1,000 °C and floating along the surface of a bath of molten tin. The surface of the tin is completely flat and so the bottom glass surface as well as the top surface is also flat. The ribbon is then cooled and passed through an annealing oven. The result is glass of uniform thickness and with a bright, fire-polished surface. The process is cheap and effective, particularly as it is a continuous surface, and variations in the width of the glass can easily be carried out.

Pilkington Brothers was (and is) the major British glass manufacturer. Alistair Pilkington was a member of the family which controlled the private company. Born in 1920, a mechanical engineer, he thought of the basic concept. The first 'pilot' production plant started in May 1957. Poor-quality glass was produced for 14 months. Then suddenly good-quality glass began, and continued, to appear. The process was publicised. Now it was decided to replace the worn-out equipment and to the amazement of everyone poor-quality glass again came out from the plant. It took three months of investigation to realise that the good glass had only occurred because of a broken part. Once the set-up was reproduced, with the broken part, good-quality glass again began to be produced.

Another piece of good fortune was that the glass as originally produced was the right thickness for half of the market. Glass was produced for sale while research was carried out into how to make thinner or thicker glass (this is done by controlling the speed with which the glass ribbon is drawn off from the furnace). The process took £7 million to develop, a huge sum for the time. Pilkington Brothers were able to license it to many foreign countries during the 1960s. Today 90% of all plate glass is made in this way.

769692
2 SHEETS

PROVISIONAL SPECIFICATION
This drawing is a reproduction of
the Original on a reduced scale
Sheet 1

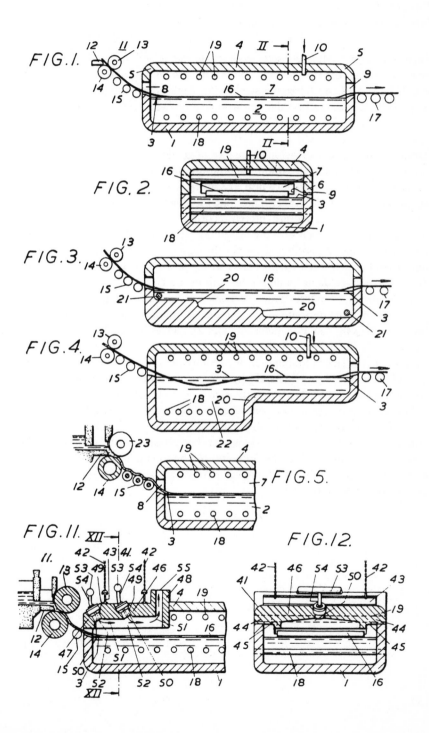

FIG.1.

FIG.2.

FIG.3.

FIG.4.

FIG.5.

FIG.11.

FIG.12.

The geodesic dome

Dome with great structural strength
Richard Buckminster Fuller, Forest Hills, New York
Filed 12 December 1951 and published as US 2682235

Fuller was born in Massachusetts in 1895. He was an architect and engineer with an early concern for the environment and making the most of resources. During World War II he was asked to design assembly-line built housing to provide employment after the war for aircraft production workers in Wichita. He came up with the Dymaxion house, a round, metal house built round a mast. An important aspect of it was itself patented as US 2351419. In working on the problems of improving the stability of the mast he developed the idea of geodesic domes.

Spheres enclose the maximum volume for the minimum surface. This can be valuable when trying to reduce heat loss, or to reduce wind loss, apart from saving on materials. 'Geodesic' is a term meaning the shortest distance between two points. Such a dome uses a pattern of tetrahedrons and octahedrons—in effect self-bracing triangles—which give the maximum structural advantage within a sphere. The load-bearing is distributed evenly throughout the structure so that it has no limiting dimension, each part being as strong as any other. There is no structural reason why a huge dome could not be built over a city.

As the size of a geodesic dome increases it gets stronger, lighter and cheaper per unit of volume, the opposite of conventional buildings. The first dome built to his design, in Hawaii, had a diameter of about 45 metres. It weighed about 30 tonnes. (St Peter's in Rome has a similar diameter but weighs 30,000 tonnes.) Further advantages are the lack of foundations as a requirement, the ease and rapidity of building and their ability to withstand earthquakes. And yet the dome is so thin that if the contents were a chicken's egg, the skin would be thinner than the eggshell.

The geodesic design was known before and was used for example by Barnes Wallis, the British engineer, for the fuselage of the Wellington bomber with his British patent GB 429188 in 1935. The design has been used in many contexts, including in hostile conditions such as the Antarctic and on mountain summits. The most famous example is probably the American pavilion at the 1967 Montreal Expo. Fuller died in 1983 in Los Angeles.

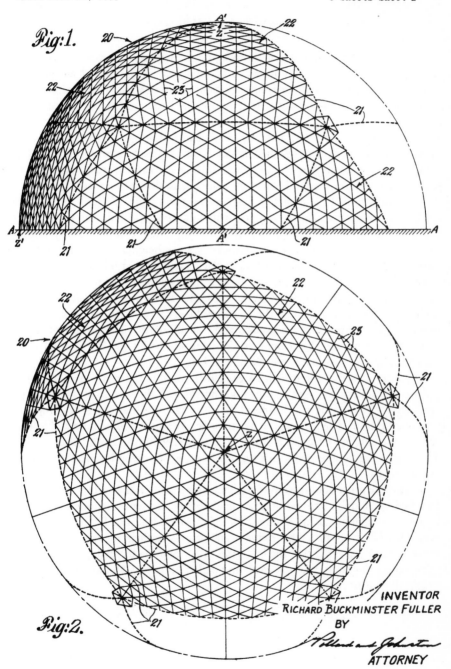

Fig:1.

Fig:2.

INVENTOR
RICHARD BUCKMINSTER FULLER
BY

ATTORNEY

The hovercraft

Vehicle moving above a cushion of pressurised air
Christopher Sydney Cockerell for Hovercraft Development Ltd, London, England
Filed 12 December 1955 and published as **GB 854211**

Cockerell was an electrical engineer working for Marconi who went off to Suffolk to become a boat builder. He began to be interested in the idea of making ships faster by reducing friction. He wondered about air lubrication, where a thin layer of air was between the hull and the water's surface. It became apparent that a thick layer was needed. He took a pair of tin cans, one with the top and bottom removed, the second, slightly smaller, with the closed end facing up, attached to a vacuum cleaner reversed to blow rather than to suck. Using scales, he measured the pressure of air going down the space between the walls of the cans. It was three times the pressure without the cans. This was the principle of the 'annular jet'.

The idea is to avoid obstacles by floating over them on pressurised air. Jets of air are blown downwards and are trapped with a curtain or skirt (invented by Denys Bliss) which, tapered inwards, taps most of the air. The pressure builds up until no more air can be squeezed in and the vehicle rises to a height of 23 centimetres. The air blown downwards 'bounces' off the surface, giving an extra benefit.

During 1956 Cockerell devised a closed vortex system that recirculated the same air. He demonstrated a model at the Patent Office and at the Admiralty, which, although interested, insisted that the invention be classified as secret. After they lost interest, the invention was declassified and was offered to the government-funded National Research and Development Corporation (NRDC). They formed a subsidiary company, Hovercraft Developments Ltd.

The story goes that the British Patent Office could not give the invention to the examiner dealing with ships, as it had no contact with the surface of the water, nor could they give it to the aviation examiner, as it had no wings or motors propelling it through the air. The first experimental model cost £120,000 to build and test and crossed the English Channel in 1959. The first scheduled service began in 1962. There were predictions about this time of huge, ocean–going hovercraft but nothing happened.

Meanwhile other countries were working along similar lines. Both the US Navy and the US Marine Corps began work in about 1957 in separate programmes. The idea was continuously adapted if only for different models. The British government persuaded Cockerell to sell his rights in the patents to them for £100,000 in 1971. When it was discovered that the US Army was using the technology without paying royalties they were taken to court and ended up paying $6 million in damages. Cockerell died in 1999.

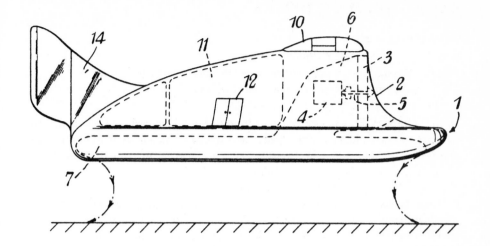

FIG.I.

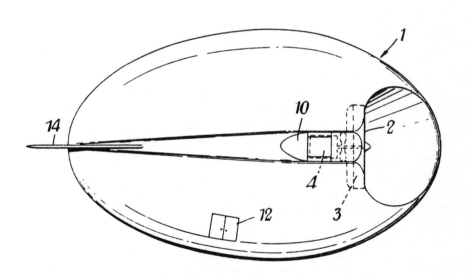

FIG.2.

Lego® building bricks

Toy building bricks
Godtfred Christiansen, Billund, Denmark
Filed 28 January 1958 and published as DK 92863, GB 935308 *and*
US 3034254 *(but all these are somewhat different from each other)*

In 1932 Ole Kirk Christiansen, a joiner from Billund, Denmark set up a small business making stepladders, ironing boards and wooden toys. He went around the district selling them door to door. From 1934 the business began to use the name 'lego', a contraction from the Danish for 'play well', LEg GOdt (it was later realised that it was also Latin for 'I study' or 'I put together').

After World War II plastics began to be used for many purposes, and in 1947 the company bought the first injection moulding machine in Denmark. In 1949 they came up with 'automatic binding bricks' which had the familiar studs at the top of the bricks but not the hollow cylinders at the bottom for them to fit into. In 1954 Godtfred, the son, was at a local toy fair. A buyer complained that everyone's products seemed to be the same and that there should be a line of toys that could fit together. He went home and made a list of ten criteria for a quality line of toys. It included compatibility between older and newer lines so that both could be used together. He decided to base the line on the automatic binding bricks. Each brick should be capable of linking with others. The Danish patent illustrated shows the hollow cylinders for the studs to squeeze between, which gives added stability, and hence is an advance on such earlier patents as Harry Page's GB 587206, filed in 1944, which resembles the automatic binding bricks idea.

From 1960 the company (by now Interlego AG, a Swiss firm) dropped wooden toys altogether and devoted all its energies to plastic building bricks and accessories. Numerous lines have been introduced as 80% of European, and 70% of American, households are thought to have some, the challenge being to encourage sales of yet more variants of the product. At first roofs, floors, wheels and so on were additional features. Trains first appeared in 1966; in 1967, Duplo® bricks were introduced, which are twice as long, wide and high and hence easier for smaller children to handle; and different figures started to appear. From 1997 Mindstorms®, bricks with software which could be programmed to carry out activities, emerged. In all cases the pieces could fit into all the other components in all the lines.

By 1998 there were 600 different sets with 2,069 elements. There were eight main lines such as space, castle and pirate themes. The company is a major exporter for Denmark. The opening of a theme park made out of Lego®, Legoland® at Billund was followed by one at Windsor, England and another at Carlsbad, California. The patents have long expired and the company has tried hard to keep out competitors by using designs and copyright. This has led to court cases, particularly in Australia and Hong Kong, with Tyco Toys, which introduced the cheaper Super Blocks in 1984, being seen as the major challenger.

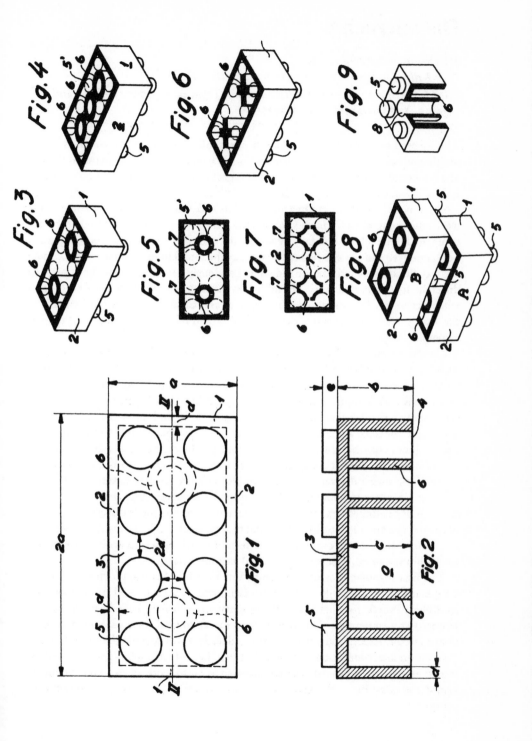

Fig.4

Fig.6

Fig.9

Fig.3

Fig.5

Fig.7

Fig.8

Fig.1

Fig.2

The microchip

Monolithic integrated circuit
Jack Kilby for Texas Instruments, both Dallas, Texas
Filed 6 February 1959 and published as US 3138743, 3138747, 3261081 *and*
3434015, *and* GB 945734 *and* 945737–49

The microchip is the basis of modern electronics because of its small size, tiny weight and reliability. It means that with a piece of semiconductor material the size of a fingernail a vast array of light and compact electronic devices can be operated. Jack Kilby was born in 1923 in Missouri. Trained as an electrical engineer, he got a job in May 1958 with Texas Instruments. His task was to help miniaturise existing electronic components. He did some work on existing components made by the company before a long holiday began for all the factory employees. As Kilby was a new employee and not entitled to the holiday, he found himself alone to think about the way things were built in the industry. Previously components had been made of different materials and were assembled by hand in conjunction with each other when necessary. He wondered why resistors and capacitors could not be made of the same material as the active devices. They could then be interconnected to form a complete circuit.

He made some sketches and showed them to his boss, Willis Adcock, on his return from the holiday. Adcock was sceptical and asked for proof that a circuit made entirely of semiconductors could work. Something was put together on an improvised basis and demonstrated in August 1958. Kilby then began work on a silicon semiconductor wafer that had existing contacts on it, and devised a circuit which used these contacts, a phase-shift oscillator. This was finished two weeks later. Figure 6a shows not what the chip actually looks like but rather is a schematic indication of how the different components work together, in this case as a multivibrator circuit. The wiring diagram for this example is Figure 6b. The basic principles laid down in US 3138743 hold true today. It provides for seven steps (nowadays usually eight) of depositing layers, heating or other activity to provide features for all or part of the microchip. The most fundamental is depositing a layer of insulation which is cut into for the electric current, the pattern of which varies according to what the microchip has to do. Transistors are used as appropriate.

The first manufacturer of microchips was Fairchild Semiconductors in 1961 in what is now known as Silicon Valley in California. This stemmed from work done by Robert Noyce for that company in early 1959, before Kilby's work was announced at a conference. On 30 July 1959 Fairchild filed for a patent for an integrated circuit. After a court battle lasting a decade it was decided that Fairchild had rights to an interconnection technique but that Texas Instruments had invented the integrated circuit. Kilby left the company in 1970 to work as an independent inventor and consultant. He holds over 60 patents and is presently working on solar energy.

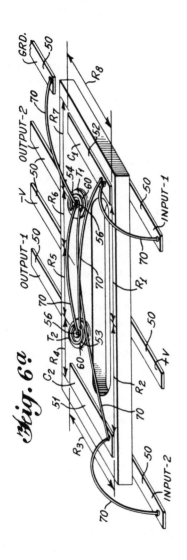

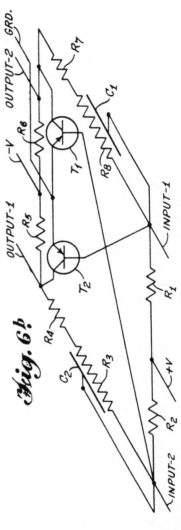

INVENTOR

Jack S. Kilby

BY

Stevens, Davis, Miller & Mosher
ATTORNEYS

The oral contraceptive

Orally taken drug which prevents pregnancy
Carl Djerassi, Birmingham, Michigan and Luis Miramontes and George
Rosenkranz, both of Mexico City, for Syntex SA, Mexico City, Mexico
Filed 5 October 1951 and published as US 2744122

Otherwise known as the Pill. Birth control advocates had long wanted a reliable form of contraception. From the 1920s it was known that hormones were involved in pregnancy and research was carried out into them. This research included making quantities of progesterone hormone from Mexican yams. Russell Marker carried out such research between 1939 and 1949, before giving up after business disputes. Marker's hormone work was carried on independently by Carl Djerassi in Mexico and by Frank Colton in Chicago. Both were born in 1923, the former in Austria and the latter in Poland. Both were thinking of using their work as a way of creating a pill to be taken orally which would help those suffering from menstrual problems. The idea that it could be a birth control device apparently did not occur to either man. Djerassi was working for Marker's company, Syntex.

About this time Katherine McCormick, heiress to the McCormick reaper fortune, was looking for a cause on which to spend money. She asked Margaret Sanger, the birth control advocate, who said that a safe and reliable oral contraceptive was the greatest need in her field. Gregory Pincus, a Massachusetts endrocinologist, was financed by her to find a contraceptive, with $2 million eventually being put into his work. Pincus worked with gynaecologist John Rock (whose main interest was in fact in promoting fertility) and together they chose Colton's later, rival product to field test from 1954 in Puerto Rico. Colton's work was patented as US 2691028 and US 2725389, which were both filed in 1953. His version was used by Colton's firm G.D. Searle and was launched as Enovid in 1960 in the United States and in 1962 in Britain.

Oral contraceptives work by preventing ovulation by using artificial syntheses of the naturally occurring progesterone and estrogen. Estrogen naturally occurs in the early part of the menstrual cycle and promotes the growth of the womb lining, while progesterone occurs in the later stage, and basically prevents ascent of the semen. In effect the body is tricked into thinking that it is already pregnant. Despite the name of the Pill there are in fact different formulations. It is now known that the early doses were far too high and that 5% of the early doses are sufficient. The oral contraceptive has been the cause of many religious and ethical objections, and many point to medical risks from heart disease and strokes for users. Others point out that some diseases are less likely to occur for users and that being pregnant and giving birth are risks in their own right. There has been some controversy between Djerassi and Colton about who deserved the credit for the Pill. Many give the credit to Pincus. The term itself was coined by Aldous Huxley in 1958 in his book *Brave New World revisited.*

1

2,744,122

Δ⁴-19-NOR-17α-ETHINYLANDROSTEN-17β-OL-3-ONE AND PROCESS

Carl Djerassi, Birmingham, Mich., and Luis Miramontes and George Rosenkranz, Mexico City, Mexico, assignors, by mesne assignments, to Syntex S. A., Mexico City, Mexico, a corporation of Mexico

No Drawing. Application November 12, 1952,
Serial No. 320,154

Claims priority, application Mexico November 22, 1951

4 Claims. (Cl. 260—397.4)

The present invention relates to cyclopentanophenanthrene derivatives and to a process for the preparation thereof.

More particularly the present invention relates to Δ⁴-19-nor-androsten-17β-ol-3-one compounds, having 17α-methyl or ethinyl substituents and to a process for producing these compounds.

In United States application of Djerassi, Rosenkranz and Miramontes, Serial Number 250,036, filed October 5, 1951, there is disclosed a novel process for the production of 19-norprogesterone. As set forth in this application, 19-norprogesterone has been found to be even stronger in its progestational effect than progesterone itself.

In accordance with the present invention, it has been found that the method described in detail in the aforementioned application may be applied to produce compounds of the androsten series, namely, Δ⁴-19-norandrosten-3,17-dione. By protecting the 3-keto group of this compound, as by the formation of a suitable enol ether as hereinafter set forth in detail and reacting the resultant 3 enol ether with suitable reagents, there may then be produced Δ⁴-19-nor-17α-methylandrosten-17β-ol-3-one or Δ⁴-19-nor-17α-ethinylandrosten-17β-ol-3-one. The first of these compounds exhibits more pronounced androgenic effects than its homologue methyltestosterone and the second of these compounds exhibits more pronounced progestational effects than its homologue ethinyltestosterone.

Certain of the novel compounds of the present invention may therefore be represented by the following structural formula:

In the above formula X is selected from the group consisting of C≡CH and CH₃.

Compounds as exemplified by the foregoing formula

2

may be produced in accordance with the process outlined by the following equation:

In the above equation R represents a lower alkyl radical, as for example methyl or ethyl, and R¹ represents a lower alkyl radical such as ethyl or methyl or a benzyl radical or any of the other groups which are customarily used as part of an enol ether customarily used for the protection of the 3-keto group of steroids. Thus, in the alternative rather than an alkyl or benzyl enol ether as shown benzyl thioenolethers may be utilized in the present reaction or other thioenolethers.

In practicing the process of the present invention, a suitable 3 lower alkyl ether as for example 3-methoxyestrone is dissolved in a suitable solvent such as anhydrous dioxane. Thereafter anhydrous liquid ammonia and an alkali metal, such as lithium or sodium metal, are added to the mechanically stirred solution. The stirring is continued for a short period, as for example one hour, and a quantity of ethanol is then added. When the reaction is complete and the blue color produced disappears, water is then added. The ammonia is then evaporated on a steam bath and the product collected with 2 l. of water. Extraction with a suitable solvent, such as ether, and ethyl acetate followed by evaporation to dryness under vacuum, produced a yellow oil. The oil thus obtained was then dissolved in a suitable solvent, such as methanol, and refluxed with a mineral acid, such as hydrochloric acid, for approximately one hour. After purification, extraction and so forth, the product obtained was a yellow oil having an ultraviolet absorption maximum characteristic of a Δ⁴-3-ketone. The last-mentioned yellow oil was then oxidized as by adding chromic acid in acetic acid to a

Scrabble®

Board game
James Brunot and Helen Brunot, Newtown, Connecticut
Filed 11 June 1954 and published as **GB 747598** *and* **US 2752158**

Scrabble® was actually devised not by the Brunots but by Alfred Butts, an architect from Poughkepsie, New York state. It was 1931, and like many newly unemployed he found that time hung heavy on his hands. He decided to devise a game which would be half luck and half skill. He read through the front pages of the *New York Times* to work out how frequent particular letters were in the English language (but reduced the occurrences for the letter S, thinking it would make the game too easy), and gave each a value based on how rare it was. No board was required, as the tiles were laid out in crossword pattern. He called it Lexico.

His patent application was rejected and games manufacturers were uninterested. In 1938 he modified the game by adding a 15 by 15 square board, with high-scoring squares, and a seven-tile rack (these features are still used today). The name changed to Criss-Crosswords. Again neither the Patent Office nor games manufacturers wanted to know. He did make a few, but otherwise returned to being an architect.

In 1948, James Brunot, owner of one of the few sets, said that he was willing to try marketing the game. In return for royalties Brunot acquired the rights. He rearranged the premium squares, simplified the rules and changed the name to Scrabble®, which was registered as a trade mark the same year, and in Britain in 1953. Working from home, he sold over 2,000 sets in 1949. By 1952 word began to get around and sales began to increase, just as Brunot was about to give up.

Jack Strauss, the chairman of Macy's, the big department store, played the game while on holiday. On his return he asked his games department to send him some but they had none in stock. Macy's began to support promotional efforts. Brunot was unable to cope with the increased sales so he licensed production to Selchow & Righter. Rights outside the United States, Canada and Australia were sold to J. W. Spears, a British company. The first patent application was as late as 1954 and was made in Britain.

Different versions are needed for different languages as the letter frequency and sometimes the letters can vary (e.g. Spanish, which has LL and CH). Over 100 million sets have been sold in 29 languages. James Brunot died in 1984 and Alfred Butts in 1993.

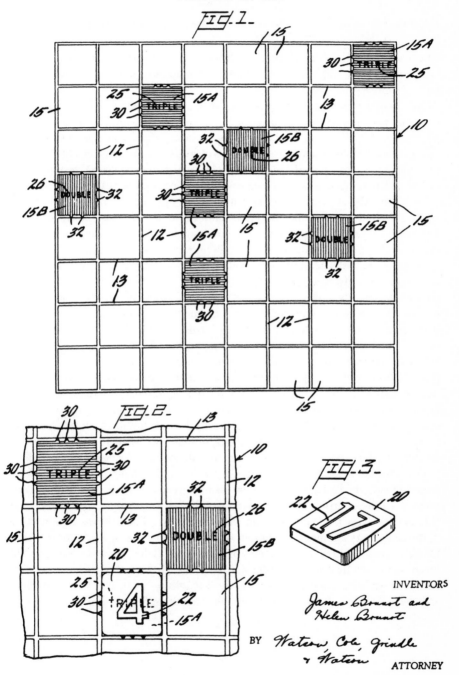

INVENTORS
James Brunot and
Helen Brunot

BY Watson, Cole, Grindle
& Watson

ATTORNEY

The shoulder-lap seat belt

Seat belt for use in vehicles, going diagonally across the body
Nils Ivar Bohlin for Volvo AB, both Goteborg, Sweden
Filed 29 August 1958 and published as US 3043625 *and* GB 870423

The invention of the three-point seat belt which goes diagonally across the body has been hailed as the single most important contribution to safety in automobile history.

Nils Bohlin was an engineer working for Svenska Aeroplan AB, a Swedish aircraft company, developing and patenting inventions to do with ejector seats. He was hired in 1958 by Volvo as their first safety engineer, whose role was to ensure that safety issues were looked into in the designs of Volvo's car models. His first task was to improve the conventional two-point seat belt which buckles across the waist. This design is dangerous in a sudden stop or collision because the head is not restrained from hitting the steering wheel or windscreen, and because of the constriction round the abdomen. It is also uncomfortable and awkward owing to the difficulty for the user reaching for anything. So from designing ways of getting people out of moving vehicles, Bohlin switched to designing ways to keep people in.

He came up with the three-point belt. Though still not used by every driver or passenger, virtually every car now has such belts. By running the belt across from one shoulder to low down on the other side more leeway is provided for reaching, making the belt more comfortable and convenient. The only major change, other than small comfort considerations, was the introduction by Volvo of the inertia reel in 1968. The improved design with a sliding buckle allows deceleration before the belt halts the user, optimising the restraining force in an accident. Volvo immediately introduced the design into their new models' front seats, and by 1963 they were in all their models. There was some resistance from manufacturers despite the fact that Volvo deliberately did not enforce its patents so as to encourage others to adopt the life-saving measure. Nobody really knows how many lives have been saved but the number is clearly enormous, despite the fact that many still do not wear belts. More recently they have been introduced into the back seats as well.

Great Britain made seat belts compulsory in 1983. The United States has left the issue to the individual states with New Hampshire alone not having made it compulsory. Both countries have had extensive publicity campaigns, the former with 'Clunk click every trip' and the latter with 'buckle up'. The belt was chosen as one of the eight most significant inventions of the past 100 years by the German Patent Office in 1985. Bohlin is now retired, having done further work on seat belts and means of protecting occupants from side collisions. He says, 'Sometimes I get a call from some grateful person who has survived thanks to the belt. It warms my heart and shows that I really have been able to do something for mankind'.

July 10, 1962   N. I. BOHLIN   3,043,625

SAFETY BELT

Filed Aug. 17, 1959

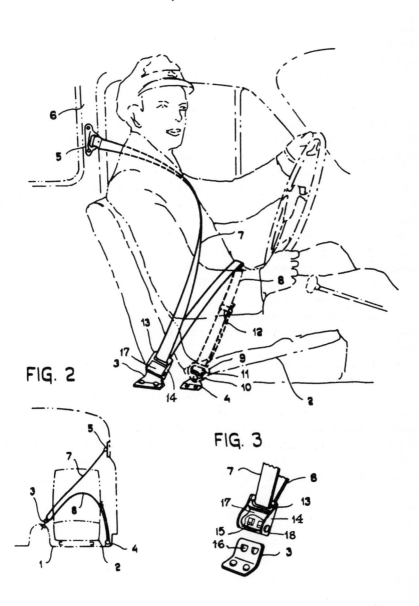

FIG 1

FIG. 2

FIG. 3

Velcro® fasteners

Nylon hook and loop fasteners
George de Mestral, Prangins, Vaud, Switzerland for Velcro SA, Fribourg,
Switzerland
Filed 22 October 1951 in Switzerland and published as GB 721338 *and*
US 2717437

This product, popularly (but erroneously) known by its trade mark, originated in dissatisfaction with jammed zippers. De Mestral was a Swiss engineer. One night in 1948 when he and his wife were about to go out to dinner, he was frustrated by a stubborn zip on her dress, and idly wondered if there were another way to secure fabrics.

A few weeks later he took his dog for a walk through the forest. On his return he noticed cockleburs on the dog's coat and thought that he would look at one under the microscope. The surface consisted of tiny hooks, and he noticed that it readily stuck to tiny loops in his clothing. He wondered if this principle—of tiny hooks and eyes—could be made into a product.

He spent eight years working on the idea. He wanted a cheap and simple way of making large quantities of the fasteners. At first cloth was thought of as the appropriate material, but nylon was later found to be better. The loops could easily be made by mass-production but the hooks were more difficult. The answer was to make the loops mechanically, then to heat them so that they became raised from the surface, and then to cut them so that the clipped ends became hooks. The result was a product which was either soft and fuzzy (loops) or rough (hooks), which were simply pressed together or pulled apart.

The trade mark Velcro® was a combination of the French words *velours croché*, hooked velvet. It was registered as a trade mark in the United States in 1958 and in Britain in 1960.

De Mestral sold the rights to Velcro SA, run by Jean Revaud, a French-born American, who began manufacturing in France. De Mestral made millions in royalties. Initially the garment industry was not interested, but its usage spread, including in the aerospace industry during the 1960s, which used it for fastening insulation in fuselages. It has also been very useful for children and the handicapped as no coordination is needed to use the product. Later improvements have included different colours besides the original black, and a quiet version for ammunition packs to avoid the detection of soldiers.

Fig.1

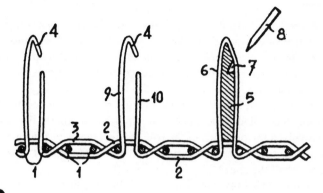

Fig.2

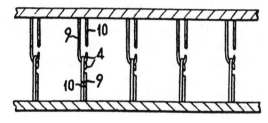

Fig.3

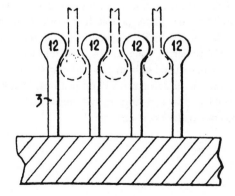

The Wankel rotary piston engine

Engine involving a rotor moving round within a chamber
Felix Wankel, Lindau, Germany for NSU Werke
Filed 7 February 1956 and published as DE 1012309, GB 791689 *and* US 2988008

Felix Wankel was born in 1902 in Lahr in the Black Forest. He did not go to university but instead worked in sales in a scientific publishing house. He lost his job, and then spent his time tinkering in his car repair workshop. From 1926 he began to think of an alternative to the internal combustion engine, as so much energy is wastefully made into heat rather than driving the car. In 1936 he received a patent for an engine which involved a rotor moving round a fixed central shaft. However, the entire engine had to be taken apart in order to replace the spark plug. A later model involved the principle of the true Wankel rotary piston engine: 'planetary rotation'. At first he thought of a casing which rotated, with an internal rotor rotating inside it at half the speed.

It was not until 1951 that Wankel was able to run his own proper workshop to develop his ideas, supported by the engineering company NSU. The first trial of the engine illustrated was not carried out until 1 February 1957, after the date of the original application. The upper drawing shows the principle of a body within a chamber moving round. It is a four-stroke cycle engine which means that there is a cycle of four occurrences. The rotor in the middle of the chamber moves around and as it does so the space between it and the chamber wall varies in size. Fuel is drawn in at one port, is compressed as the rotor moves, is ignited and is expelled at a second port. The ports are not shown. The chamber moves as well as the rotor at a ratio of 3 to 2. Later designs modified it so that instead of the chamber shown there was a chamber rather like an ellipse with a rotor shaped like a bulging triangle, and a stationary chamber was adopted. The idea of a rotary piston engine was not new but his work on the geometric shape of an enclosing chamber that was sealed and did not involve valves was new.

The design is simple but powerful. Advantages over the internal combustion engine include reduced size and weight, vibration, noise and production costs. Drawbacks have been high fuel consumption and high exhaust emissions. There is also high wear at the tips of the central rotor. Wankel had not investigated the patent literature but rather had thought up numerous different designs and tried out many of them. He found out later that many people had worked on the same ideas but their designs did not have such a high compression ratio. The Wankel engine was first used in the Wankel Spyder in 1963. In 1974 Suzuki used it in a motorcycle. The engine does not have to be used only in a car, of course, and work has been carried out on industrial, marine or aeronautical applications. Wankel himself died in 1988 but Wankel GmbH is still carrying on his work improving the engine, although the concept has not really been successful.

Abb. 3

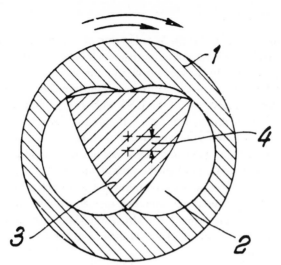

Drehzahlverhältnis
Innenläufer:Aussenläufer
2:3

Abb. 4

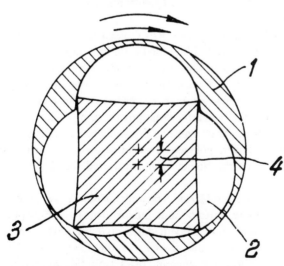

Drehzahlverhältnis
Innenläufer:Aussenläufer

3:4

1960–1969

THE generation born after World War II came of age in the 1960s. The decade became identified in the West not only with much better standards of living (with great strides made, too, by Japan's economy) but also with an increasing youth culture which held more liberal attitudes towards behaviour and belief. The Presidency of John F. Kennedy of the USA with its heralding of a New Frontier seemed an emblem of this and to be a shift towards government becoming the province of younger politicians, contrasting with the age of many world leaders elsewhere and until then. Kennedy's senseless murder would darken many people's optimism but it would not halt the pace of change of all kinds. During these years, as the USA and the Soviet Union duelled with each other for world-wide influence and in technological development, men and women were launched into space, walked in space and orbited the Moon and Mars, until eventually 600 million people watched by satellite television men from the Apollo 11 space mission walking on the Moon.

Contraceptive pills, laser technology, hip replacements, the first heart transplant and the initial development of heart by-pass surgery were to change for the better the lives of so many. So would more equal political rights even in a democracy like the USA. While Martin Luther King mobilised both black and white people's support for civil rights, black people in the USA were given greater constitutional rights and freedom by the Civil Rights and Voting Rights Acts of 1964 and 1965, masterminded by Lyndon Johnson as an essential part of his vision of a Great Society.

Between 2 and 3 million people had crossed from East into West Germany since the end of the 1940s. Building the Berlin Wall in 1961 was designed by East Germany to end this emigration. Brinkmanship between the superpowers over Berlin and, more so, in the Cuban missile crisis of the following year pushed the superpowers closer to nuclear war than ever before. But Kennedy's resolution in 1962 ushered in a period when the superpowers drew back from stark confrontation, with nuclear weapons as a mutual deterrent between them, even if they continued to pursue opposing policies on world issues and to negotiate moments of acute tension as when the Soviet Union thwarted the flowering of the Prague Spring in 1968.

'The wind of change', in Harold Macmillan's phrase, swept across Africa (and Asia) as more and more states became independent from European government. However, in some newly independent states, civil war and the splintering of communities ensued as chaos and tragedy in the Congo and Nigeria testified. Tensions between white dominance and black aspiration also triggered rioting and armed outbreaks in South Africa, Rhodesia and Angola. China's communism clashed increasingly with the Soviet Union's interests before and after Khruschev's fall. Youth culture of a different kind was promoted by Mao Zedong in the Cultural Revolution. His encouragement of young Red Guards was intended to refresh revolutionary beliefs in China as well as to bolster Mao's personal power while an ideological battle was waged between the modernisation of Chinese structures or permanent revolution for them. The Cultural Revolution led to much suffering and loss of life: it also proved an industrial and agricultural disaster.

The Arab-Israeli Six-Day War of 1967 gained Israel much Arab territory and provided another potential flashpoint for US-Soviet confrontation. To roll back communist influence in eastern Asia the USA was drawn into fierce conflict in Vietnam. This brought further progress to military technology but also showed that the technology, though it might devastate city and country, could not defeat by itself opponents who could vanish by day and attack by night. The effect of the Vietnam war on American society became destructive and divided those who saw the war's purpose as right, and victory essential, from those who saw its purpose as wrong and its conduct damaging in the eyes of world democracy. It questioned the values of the superpowers' authority and peacekeeping abilities that until then had received too little public challenge. Some of the political aspirations of the free world that had accompanied Kennedy's election in 1960 were in tatters by 1969 but the Moon landing showed how far science and technology had travelled in 10 years.

Two significant achievements of construction and architecture lay in Egypt's High Aswan Dam and in Brasilia's inauguration as Brazil's new capital. Transport technology introduced the 'bullet' trains of Japan, the first hovercraft service across the English Channel, the initial flights of the Anglo-French supersonic Concorde airliner and the Boeing 747 jumbo jet, and the vertical take-off and landing Hawker Harrier 'jump jet' attack aircraft. The Telstar communications satellite enabled live TV to be broadcast across the world, the first weather satellite was launched, the fibre-optic telephone cable invented and industrial robots began to be used in factories. Measles vaccine was perfected but deformed babies caused by their mothers' use of thalidomide showed the danger which man-made drugs could bring. Rachel Carson warned of the hazard of chemical pesticides, the link between cigarette smoking and lung cancer was confirmed and there was a first intimation of the 'greenhouse effect' which could raise temperatures and sea levels owing to atmospheric carbon dioxide generated by human activity. Two 1969 developments were to have a great impact on the rest of the century. The US Department of Defense established the Internet. At the opposite and smaller end of the information technology revolution Engelbart invented the 'mouse' computer input controller.

Computer-aided tomography

Electronic scanning of the body or parts of it
Godfrey Hounsfield for EMI Ltd, Hayes, London, England
Filed 23 August 1968 and published as GB 1283915 *and* US 3778614

Godfrey Hounsfield was born in 1919 in Nottinghamshire and grew up on a farm, where he loved making up his own scientific experiments. He worked in the radar field during World War II and then went to college. From 1951 he worked for EMI Ltd, an electronics company, specialising in the new field of computers. After a project turned out not to be viable, Hounsfield was given the opportunity to think in–depth about what area he wanted to research next. One of the ideas he suggested was enabling computers to recognise patterns so that they could 'read' letters or numbers. While walking in the countryside, he thought of merging this idea of pattern recognition and the principles of radar. An X-ray machine would make an image of thin slices of a patient and a computer would process the data to produce a three-dimensional image. Conventional X-rays were useful, but did not indicate how deep in the body the problem was.

In 1967 such a machine was not practicable as computers were not sufficiently far advanced. Initial machines took 2.5 hours to scan an object followed by 2 hours to convert it to an image. After much work on prototypes, which included such experiences as Hounsfield himself having to take a bullock's brains across London by public transport, a machine was built in 1971 which could scan the brain. It went on sale the following year. It works by an X-ray beam rotating 180 degrees around the head creating projections in each view, which are combined for the total picture. This method was far safer than the normal activity of pumping chemicals into the brain and detecting them. By 1975 improved models enabled the entire body to be scanned with a moveable table taking the patient into a circular hole in the machine. Figure 1 shows a conventional X-ray of a body 1, a bone 2, and a tumour 3. Figure 2a shows the same body, bone and tumour with the X-ray source at 6 and the detector at 7. This scanning technique is called tomography and the 'CAT' or 'body' scanner was hailed as the biggest advance in diagnosis since X-rays themselves.

Body scanners were expensive (£1 million each) but when the costs of using other methods or of not being able readily to diagnose the patient were taken into account they were cheap and effective. For years there were well-publicised appeals to raise money for local hospitals in Britain to acquire a body scanner. Hounsfield received the Nobel Prize for medicine in 1979 jointly with the American Allan Cormack, who had been working on theoretical models of the same idea since about 1957. EMI, which had financed the development of the scanner from its music interests, was disappointed in its sales and sold out its interests to General Electric in 1980.

1283915   COMPLETE SPECIFICATION

5 SHEETS   This drawing is a reproduction of
the Original on a reduced scale
Sheet 1

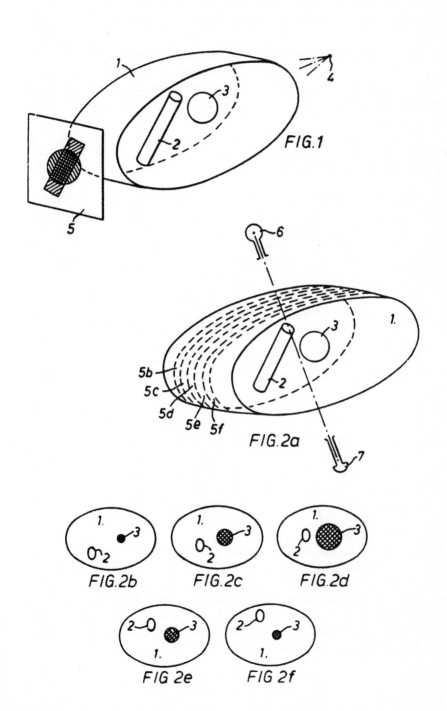

FIG.1

FIG.2a

FIG.2b

FIG.2c

FIG.2d

FIG 2e

FIG 2f

Implantable pacemaker

Device for permanently regulating the heartbeat
Wilson Greatbatch, Clarence, New York, for Wilson Greatbatch, Inc.
Filed 22 July 1960 and published as US 3057356

The realisation that, to a considerable extent, surgery can be carried out in the heart has been relatively recent. It used to be thought that any intervention resulted in death. An important problem has been dealing with 'heart block', where the signal to beat does not get through to the ventricle of the heart. The term 'pacemaker' comes from the sinus node which controls the heartbeat with 72 electro-chemical nerve signals per minute, and which is sometimes called a pacemaker. A safety mechanism does mean an occasional beat is generated but it is hardly ideal and in people with Stokes-Adams Syndrome this does not kick in and brain damage and death occurs.

Paul Zoll, a Boston cardiologist, invented the first pacemaker in 1952 although he did not patent it. A pacemaker detects when the heart does not beat and electrically stimulates a beat. It was attached to the chest and worked by sending shocks through the chest muscles. This was both painful (it resulted in burns after a day or so) and meant that the patient had to stay wired to the equipment. It was the invention of the transistor that made an implanted pacemaker possible, as such a device obviously had to be small. The amount of power required was also important. In 1958 Ake Senning and Rune Elmquist of Sweden invented the first implantable pacemaker, which was successfully used. It was however not self-contained, as a cord ran to a powerpoint. It is also unhygienic to have wires going through skin. It worked by using electronic circuits, which may seem obvious to us today, but other researchers were working on concepts such as using radio pulses or radioactive isotopes.

Wilson Greatbatch was working his way through an electrical engineering degree at Cornell University. While working on the university's animal behaviour farm, he got talking to two brain surgeons. They explained about the problems of heart block and Greatbatch realised that an artificial pacemaker would solve the problem. Transistors did not become readily available until the later 1950s and after talking to William Chardack of the Veterans Administration, Greatbatch went home and designed a pacemaker in 3 weeks. A major problem, as shown in trials in animals, was sealing it against fluids and rejection by the body and experimental models were cast in epoxy (now they are in metal).

Figure 3 shows the wafer-like pacemaker 11, which is 6 centimetres across and weighs 113 grammes. It was designed to be under the skin so that batteries could be replaced. The same figure's 13 shows the electrode which is inserted in the ventricle of the heart. It is powered by ten mercury zinc cells and uses two transistors. Greatbatch continued to work on improving the pacemaker but also worked in fields such as biomass, plant genetics and treatments for diseases. Installing a pacemaker is now almost a routine operation.

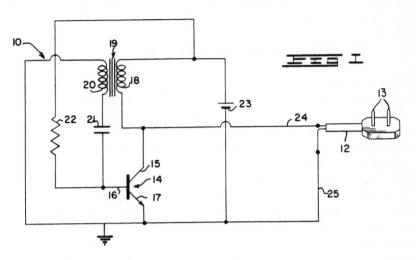

FIG 1

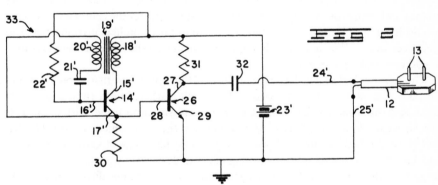

FIG 2

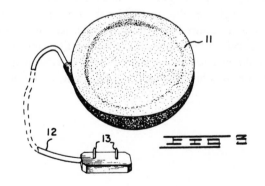

FIG 3

INVENTOR
WILSON   GREATBATCH

BY   Harmon & Kurz

ATTORNEY

Kevlar®

High-strength polymer
Stephanie Louise Kwolek, Wilmington, Delaware and Paul Winthrop Morgan,
West Chester, Pennsylvania for E. I. Du Pont de Nemours, Wilmington,
Delaware
Filed 25 April 1963 and published as US 3287323

Stephanie Kwolek was born in 1923 in Pennsylvania. After getting a degree in chemistry she meant to be a doctor, and took a job with Du Pont to earn some money for medical school. She found the work so interesting that she stayed on. Her area was textile fibres, building on Du Pont's recent work with nylon and similar polymers, which are long chains of molecules. Her work in the Pioneering Research Laboratory was supported by Hale Charch, its head, but she faced hostility from male colleagues. Women scientists were particularly rare then.

Kwolek studied the monomers which were used to synthesise a substance called PBA, or polybenzamide. They were liable to degrade easily in reaction to water. She managed to find the right conditions for them to become polymers themselves by using a solvent at low temperatures. The result was a cloudy solution, which did not sound promising, as a clear solution would have been expected. Instead of throwing it away and doing something else Kwolek tried spinning out the solution to see what would happen. The result was thin, very strong fibres, unlike anything that had been seen before. The new branch of synthetic materials she had discovered is called liquid crystalline polymers.

Du Pont worked to develop the new fibre further and found that it was five times stronger than steel of the same weight. It was not marketed until 1971, many years later, and the name Kevlar® was registered as an American trade mark in 1974. It has been used in many different applications, but the most famous is probably that of lightweight body armour. This is worn routinely like an undergarment and is not obvious. It works by the fibres behaving like a spider's web: when a bullet hits, the strands in the way of the bullet elongate and cause others not directly touched to move as well. All this helps to dissipate the energy of the bullet. Even so, ten layers or more are needed in such vests. Nylon had been used by the US Army during World War II but, while effective, it required very bulky jackets. Kevlar® is also used in such applications as skis, bridge cables, brake pads, fishing lines and tyres. The sales are in hundreds of millions annually. Kwolek retired in 1986 with 17 patents to her name. She continues to work part-time for Du Pont, encouraging women scientists.

1

3,287,323
PROCESS FOR THE PRODUCTION OF A HIGHLY ORIENTABLE, CRYSTALLIZABLE, FILAMENT-FORMING POLYAMIDE
Stephanie Louise Kwolek, Wilmington, Del., and Paul Winthrop Morgan, West Chester, Pa., assignors to E. I. du Pont de Nemours and Company, Wilmington, Del., a corporation of Delaware
No Drawing. Filed Apr. 25, 1963, Ser. No. 275,506
1 Claim. (Cl. 260—78)

This invention relates to a novel class of crystalline, linear condensation polyamides and to a manufacturing process therefor.

Among the numerous combinations of bifunctional complementary reactants heretofore suggested for the manufacture of linear condensation polyamides, those of wholly aliphatic reactants, e.g. adipic acid and hexamethylene diamine which yield nylon 66, have been most frequently employed in commercial practice since crystalline products have been readily obtained. Such crystalline products have been particularly attractive since, in comparison with non-crystalline or amorphous forms, they possess higher softening points, distinct melting points, greater light stability and opacity as well as improved tenacity and elongation values.

Although the use of aromatic polyamide reactants is also known, i.e. to achieve other desired benefits, certain combinations which include such reactants have not previously produced suitable crystalline polymers. Illustrative of such combinations are those of meta-phenylene diamine and either adipic, suberic or sebacic acid, as described for example in Flory U.S. Patent 2,244,192. Rather than exhibiting the properties of crystalline polymers, the products of those materials are amber colored resinous materials which soften at temperatures considerably below their melting point.

In accordance with the present invention, a novel class of crystalline polyamides is provided by means of a unique series of critical processing steps. In contrast to prior art polyamides prepared to contain the same recurring units, the products of the invention exhibit the desirable features which are generally characteristic of crystalline polymers and in addition afford outstandingly higher modulus values. As a consequence of these properties, fabrics produced of the novel polymers in filament form exhibit improved fabric aesthetics.

More particularly the invention provides a linear polyamide selected from the group consisting of polymetaphenylene adipamide, polymetaphenylene suberamide, and polymetaphenylene sebacamide, said polyamide having an inherent viscosity of at least 0.8 and a crystallinity index above 50.

The novel crystalline polyamides of the invention consist of recurring structural units of the formula:

$$-\overset{H}{\underset{|}{N}}-\underset{}{\bigcirc}-\overset{H}{\underset{|}{N}}-\overset{O}{\underset{||}{C}}-(CH_2)_m-\overset{O}{\underset{||}{C}}-$$

wherein m is an even number integer from 4 through 8. In filament form, the polyamides have excellent light stability and improved tensile properties, particularly in terms of higher modulus and tenacity values.

The method of the invention involves the following series of critical steps to obtain a polyamide in a highly orientable, crystallizable form:

(1) Forming an anhydrous condensation mixture consisting of (a) a diacid chloride of an acid selected from the group consisting of adipic acid, suberic acid, and sebacic acid, together with (b) a stoichiometric equiva-

2

lent, based on the said diacid chloride, of meta-phenylene diamine, and (c) an organic, acid accepting diluent which is a solvent for said diamine and said polyamide and which is present in such an amount to provide a final concentration of 5–35% by weight of polyamide therein, by addition of said diacid chloride to a solution of said diamine in said diluent while maintaining the said condensation mixture under continuous agitation in the absence of water at a temperature between −20 and 25° C., preferably 0–25° C., until completion of said addition,

(2) Continuing said agitation at a temperature not above 30° C. until a polyamide is formed having an inherent viscosity of at least 0.8, and

(3) Separating the polyamide thusly formed.

Strict adherence to the foregoing process variables is necessary to obtain the particular polyamides in such form that they be capable of crystallization to a crystallinity index of 50 or more. The reactants and diluent should be highly purified such that water and other extraneous materials are essentially excluded. The water content of the reaction mixture, in particular, should not exceed 0.02% by weight. For purposes of excluding moisture and other reactive vapors, the reaction should be conducted under an inert atmosphere such as nitrogen, dry air or the like. The stoichiometric equivalent of diacid chloride in undiluted form should be added to the diluent solution of diamine as rapidly as possible but without permitting the temperature to exceed 25° C. during the addition. For this purpose it is normally desirable to initially cool the diamine solution to a temperature between 0 and 10° C. Even after completion of the addition, the reaction mixture should not be heated above 30° C.

Following complete addition of the diacid chloride to the amine solution it is desirable to allow the reaction mixture to warm up to between 10° and 25° C. while continuing the agitation for at least 20 minutes. Subsequently the polycondensate can be precipitated by pouring the reaction mixture into water under conditions of agitation. The recovered, washed and dried polymer is then desirably dry spun from dimethylacetamide or dimethylsulfoxide to give fibers with excellent tenacity and elongation properties.

Suitable organic, acid accepting diluents in which to conduct the polymerization include alkyl substituted amides such as N-methylpyrrolidone, dimethylacetamide and hexamethyl phosphoramide.

The manner in which the prepared polyamides are crystallized is not a critical feature of the invention. Although steam setting of drawn fibers, as performed in the examples, is a preferred technique, others are also suitable. Typical among such other methods are those involving dry-heat setting, hot-wet treatment, or the use of hydroxyl-containing organic non-solvent swelling agents, such as described in U.S. Patents 2,307,846 and 2,289,377.

It is to be understood that once the polymers have been prepared in a highly purified crystallizable form, they can be suitably modified as by the inclusion of dyes, fillers, delusterants, anti-static agents and the like.

The crystalline polyamides can be employed in the form of filaments, sheets, rods, tubes, coatings and the like. In the form of filaments they are particularly useful in the fabrication of tire cords, carpets, wearing apparel, etc.

Examples I through III hereinafter illustrate the preparation of polyamides according to the process of the invention. Example IV illustrates the crystallization of those polyamides in the form of filaments. Example V demonstrates comparative properties of polymetaphenylene adipamide versus polyhexamethylene adipamide. In the examples, all parts are by weight unless otherwise stated.

The Lava-lamp®

Globules of oil and wax rising and changing under heat within a clear receptacle
David George Smith for Crestworth Ltd, Poole, Dorset, England
Filed 18 March 1964 and published as GB 1034255

Craven Walker had left the Royal Air Force at the end of World War II. He came across a primitive device in a Hampshire public house—'a contraption made out of a cocktail shaker, old tins and things'. Its creator, a Mr Dunnett, was dead. Walker purchased it and over the next 15 years worked on making a better version, while also making naturist films and running a house-swopping agency. He formed his own company, Crestworth, to market the invention.

The basic idea is that there is an attractive display in a transparent container of globules (which consist of a mixture of oil and wax) moving upwards within water as a result of heat from a light bulb in the base. The globules change shape and separate and join up again, gradually sinking as they cool to carry on the cycle. The light bulb of course provides light as well as heat for the lamp. The globules can be in one of a number of colours (derived from dyes), while sometimes the water medium itself has a different dye. The recipe of oil, wax and 'other solids' is still secret but includes paraffin and mineral oil, and becomes fluent (non-solid) at about 40 to 50 °C. Originally marketed with a large gold base with tiny holes to simulate starlight, the product was popular in the days of psychedelic music and continues to sell well in a variety of designs. Walker himself said, 'If you buy my lamp you won't need drugs'.

The initial patent was credited to David Smith but this was supplemented by Walker's GB 1186679, filed 3 years later. It states that 'some viewers may become impatient during the preliminary heating up period'; that if the lamp were left switched on for a long period the globules might split up into tiny droplets because of overheating; in cold climates the water medium could freeze; and that the lamp needed to be carefully positioned to prevent draughts etc. The patent claimed to overcome all these problems by mixing the water medium with 'an additive of a water-miscible liquid effective to raise the coefficient of cubic thermal expansion of the water'. In effect, anti-freeze such as glycol was added.

The lamp was marketed in Europe as the Astro lamp. Two American entrepreneurs saw it at a German trade show in 1965 and bought the North American manufacturing rights. It continues to be made in Chicago as the Lava Lite® lamp. In 1990 Walker was bought out by Cressida Granger with her Mathmos company. Over 7 million lamps had been sold but sales had declined. Sales have increased again to 800,000 units in 1998 with nearly half being exported, mainly to Germany. The company has won a Queen's Award for Exports.

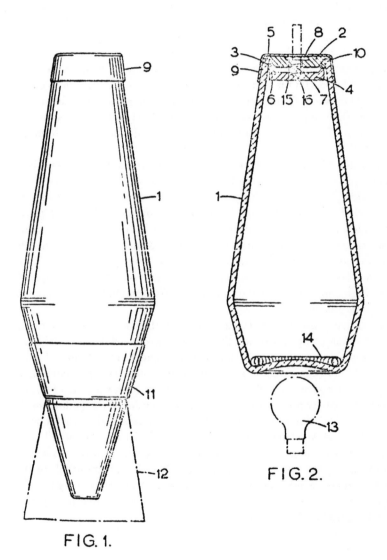

FIG. 1.

FIG. 2.

The Maclaren® baby buggy

Lightweight, foldable push-chair or stroller
Owen Finlay Maclaren, Barby, Warwickshire, England
Filed 20 July 1965 and published as GB 1154362

Owen Finlay Maclaren was a retired aeronautical designer and test pilot. His career included designing the undercarriage of the Spitfire. He applied this background to the design of the first easy-to-use push-chair or stroller. Previously, parents had to lug heavy and non-foldable contraptions around. His new push-chair only weighed 3 kilogrammes as it was made of aluminium, a light yet strong metal, and its folding mechanism meant that, when for example getting on a bus, it could easily fold across the seat. It can be folded with one arm while the other is holding the baby: the brake is kicked loose, and the spare arm folds across the push-chair, folds down, and picks it up. Maclaren referred to this concept in the patent as 'stick folding'. A design director at the company later said, 'He solved a very difficult three-dimensional fold problem. Today we are using computer-aided design systems to model solutions like that'.

Maclaren founded Maclaren Ltd which initally began production in his own converted stables, selling 1,000 units in the first year. Moves were made to new premises in Long Buckby, Northamptonshire as sales gradually expanded to over 500,000 a year, with nearly half being exported. Continual improvements have been made. This includes allowing adjustable positions for the back as babies under 6 months old should not sit upright; 'balloon' foam tires which provided a smoother ride (the processing and production of which was worked out in cooperation with ICI); and front swivel wheels and linked brakes for safety, introduced in 1981. The company's products have been commended by the Consumers' Association for the safety features built into the models.

Variant models have also been made, such as one for lightweight invalids; the E-type, in 1983, its best-selling model; and the Duette, introduced in 1991, which accommodated two babies in adjacent seats, and was designed to be capable of going through a standard doorway. Attention to detail includes providing for shopping to be carried by the push-chair as well as rain hoods. It has been suggested by some that the bigger and heavier models somewhat negate the value of the folding mechanism, as they are more clumsy to handle. Maclaren himself died in 1978 but his work goes on. In 1992 Maclaren was chosen as the official supplier to all three Disney resorts. Low-cost imports have been a problem to the company, so models called the Imp, the Scamp, the Minx and the Pixie have been introduced for this market. Recent patents include improving the grip of the handles by using two different materials and providing for a disc where a seat inclination used for a particular child can be stored in the 'memory'.

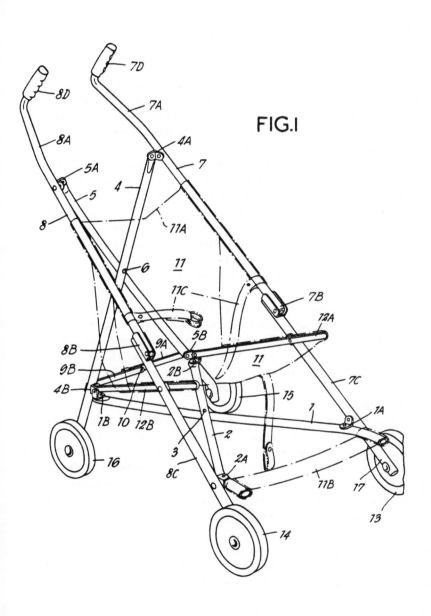

FIG.I

The mouse

Device for controlling operations on a computer screen
Douglas Engelbart, Palo Alto, for Stanford Research Institute, Menlo Park,
both California
Filed 21 June 1967 and published as US 3541541

Douglas Engelbart was born in Oregon. He became an electrical engineer, but had been a radio technician in World War II. He worked for the forerunner of NASA but began to feel in the early 1950s that he ought to be doing something, as an engineer, which would help solve world problems. He took an interest in the early articles about computers and began to imagine 'flying' through information on a screen. There were no computer science departments around so he did a PhD at Berkeley and began teaching there. He was still enthusing about using computers until another professor tipped him off that he was unlikely to gain promotion if he continued to predict that they would be important.

He switched to working for rival Stanford. From 1959 he had his own laboratory, later called the Augmentation Research Center. Here he and others worked on developing a framework for using skills and knowledge with computers. He realised that the rate of change and urgency was leaping rapidly forward and that tools were needed to cope with it. He contributed to many ideas that are common today such as the concept of windows on a screen, teleconferencing, hypermedia, groupware, e-mail and the Internet. He linked most of these ideas together in 1968 with a 90-minute demonstration of a network. Software was not then patentable but he nevertheless has over 40 patents

The invention of the mouse was one small part of the research carried out at the laboratory. Engelbart wanted an easy way to control things happening on the screen. A keyboard could not do, or not easily do, many things which someone might want to do, such as to draw. He listed all the devices currently available, or which had been suggested, such as light pens and joysticks. The best points of each were listed, and the combined tool turned out to be the mouse. It was immediately nicknamed that because it looked like one.

The first mouse was slightly different from current models as it had the cord attached to the side nearest the user rather than away, but that was swiftly changed when it was realised that it was inconvenient. Another difference was that instead of a ball at the bottom there were two discs, one parallel to the length of the mouse and one at right angles to it. These controlled the 'X–Y position indicator' as the mouse's movement controls movement up or down the screen. Most modern mice have the same discs but hidden within the casing. Right and left clicking came in later.

There is a legend that Steve Jobs of Apple invented the mouse but in fact his Lisa, later the Macintosh, was the first time it had been used commercially. He is believed to have paid $40,000 for the rights. Engelbart now runs the non-profit Bootstrap Institute in Fremont, California which promotes 'collective IQ', that is, how quickly a company responds to a situation.

Nov. 17, 1970 D. C. ENGELBART **3,541,541**

X-Y POSITION INDICATOR FOR A DISPLAY SYSTEM

Filed June 21, 1967 3 Sheets-Sheet 1

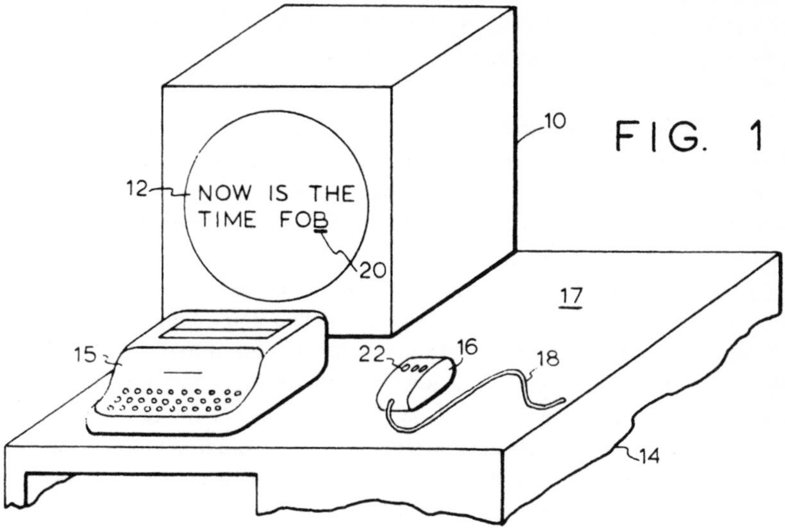

FIG. 1

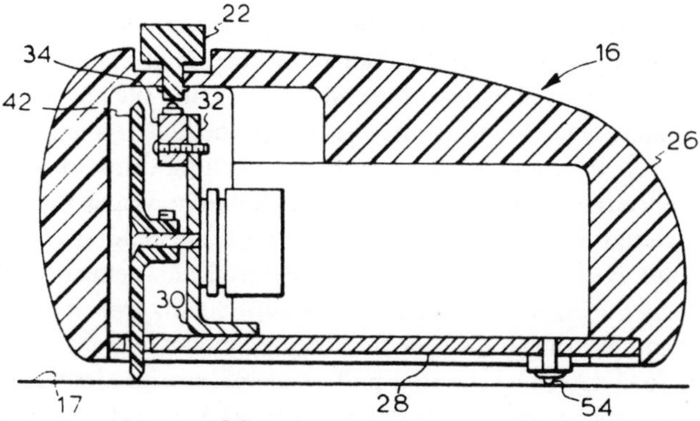

FIG. 2

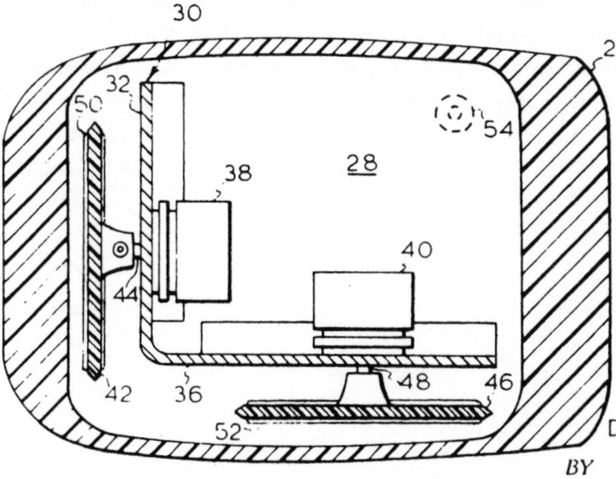

FIG. 3

INVENTOR.
DOUGLAS C. ENGELBART

BY Lindenberg + Freilich

ATTORNEYS

The ring-pull can

Drink can with a self-contained ring-pull or pull-top for opening
Omar Brown and Don Peters, Dayton, Ohio for Ermal Fraze
Filed 6 July 1965 and published as US 3349949

Ermal or 'Ernie' Fraze set up his own tool-making company, Dayton Reliable Tool & Manufacturing Company, or DRT, in 1949 in Dayton, Ohio. One day in 1959 he was at a picnic and wanted to open a drink can. Nobody had a can opener. This is not the one which cuts round the rim of a tin plate can but rather the device with a triangular cutting tool ('church key') at one end which was used to make an opening at each side of the top of the can (both sides, because if you make one opening only you get the drink up your nostrils: try it if you don't believe it). So he used a car bumper to get the can open, and the result was a lot of foam.

He said, 'There must be a better way': a method of opening integrated with the can itself while keeping the contents intact and hygienic. Drink cans were beginning to be made of aluminium instead of tin plate so he cooperated with Alcoa, a manufacturer of such cans. One night he drank too much coffee and couldn't sleep. Fraze went down to his workshop and tinkered with his tools. By dawn he had worked out the basic principles of the ring-pull can. He did have the advantage of being skilled with metal forming and scoring from his work. The idea sounds so simple. A ring that can function as a lever is placed just above the level of the top of the can, and is riveted to a prescored strip. By pulling on it the seal is broken and the tab comes away. The hole is deliberately made big to ensure that not all the hole is filled with drink when in use, otherwise the same problem as above of getting drink up your nose occurs. The patent encourages a firm hold of the ring-pull to prevent accidents and suggests a ring-pull designed to provide a secure grip for small children.

The first commercial use was by Iron City Brewery, a Pennsylvania company. At first there was resistance by the major American brewers until Schlitz agreed to use it. Gradually its use spread until ring-pulls were the accepted way of opening drink cans. *Time Magazine* in 1998 described the ring-pull as one of 'one hundred great things'. The classic 'push-in fold-back' was introduced in 1977. A problem was that the detached ring-pulls were sometimes used to jam parking meters, in the days when a broken meter meant no payment for parking (see page 94). More seriously, the introduction of the non-removable ring-pull at the end of the 1970s by the Continental Can Company meant less litter and removed danger to bare feet and to pets. Non-removable ring-pulls are compulsory in Australia and the United States but not in Britain.

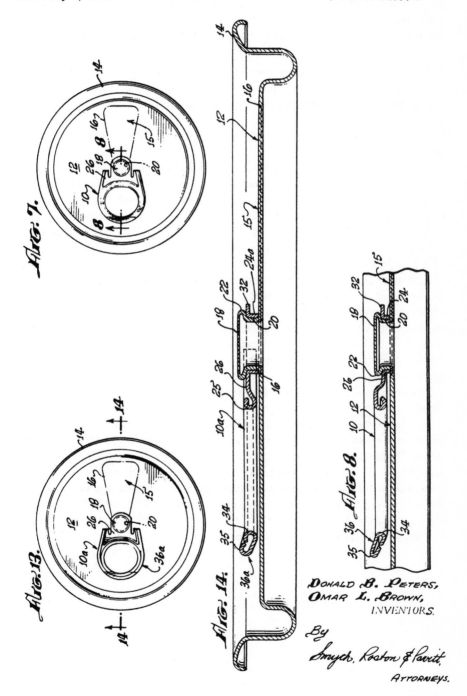

DONALD B. PETERS,
OMAR L. BROWN,
INVENTORS.

By

Smych, Roston & Pavitt,
ATTORNEYS.

The sailboard

The so-called 'Windsurfer®'
Hoyle Schweitzer, Pacific Palisades and James Drake, Santa Monica, both
California
Filed 27 March 1968 and published as US 3487800 *and* GB 1258317

Hoyle Schweitzer was a businessman who enjoyed surfing and James Drake was an aeronautical designer. Drake thought of the idea of adding a sail to a surfboard while driving home one day, and they developed the idea together. Broadly speaking, an articulated mast is attached by a universal joint to the board while a boom running horizontally is used to give direction. The boom also enables the user to hang on and no rudder is needed.

Originally fibreglass was used for the board, but the cheaper and longer-lasting polyethylene was later used. Du Pont was so pleased by the novel use that they wrote an article about the sailboard, which was an important factor in increasing sales. In 1973 Schweitzer bought out Drake. Windsurfing International Inc. was created to sell the product, and the name Windsurfer® was registered as a trade mark in the United States in 1974.

A British patent, GB 1258317, was also obtained, and Windsurfing International alleged infringement by Tabur Marine, so a court case was heard at the High Court in London in April 1982. The defence was that the patent was not truly novel. Evidence was given that S. Newman Darby had thought of it earlier, although he did not patent it. A Pennsylvania inventor, he devised a model in 1964 and wrote an article about 'sailboarding' in the magazine *Popular science monthly*. The article was reprinted in Britain in 1966 in another journal. Expert witnesses were called who spoke about whether or not they would have regarded the Schweitzer/Drake model as similar to the Darby version if they had known of it from the British article at the time.

Another witness was Peter Chilvers. He said that as a 12-year-old in 1958 he had improvised a somewhat similar model and had sailed it back and forth off Hayling Island, Hampshire. As this was a public act it was admitted as evidence of prior art. Ironically, Chilvers could himself have patented his invention without trouble. The judge held for the defence and the patent was revoked after going to the Court of Appeal in November 1983 and January 1984. There was a similar Canadian case involving Trilantic Corporation, where the Darby evidence was also put forward. Windsurfing International won that case.

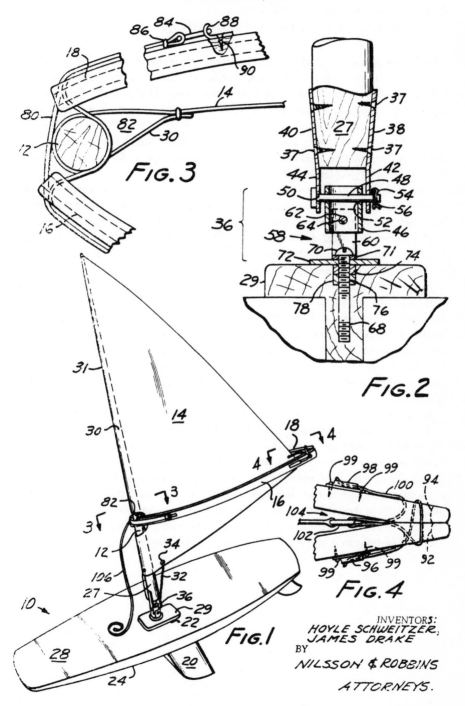

FIG. 3

FIG. 2

FIG. 1

FIG. 4

INVENTORS:
HOYLE SCHWEITZER,
JAMES DRAKE
BY
NILSSON & ROBBINS

ATTORNEYS.

The waterbed

Bed containing fluid
Charles Hall, Muir Beach, California
Filed 24 January 1969 and published as US 3585356

This has been claimed as the original waterbed, something often talked about but which few perhaps have actually used. It is described as a 'flexible bladder substantially filled with a liquid', held in place by a framework. Hall states in the patent that he is aware of previous attempts but feels that water has a tendency to transmit waves rapidly. If you turned over, you might well find yourself flung off the bed. His suggestion was that by using styrofoam or other materials to 'dampen shock waves' in the liquid, as well as to provide additional support for anyone resting on the bladder, this problem would be overcome.

The bladder itself was to be of vinyl plastic or a similar inelastic material. Hall pointed out that the bladder should not be filled too much or it would be in tension. 909 litres of water was suggested for a king size bed. The stryrofoam was in small blocks within the water. Hall suggested that hospitals would find his bed convenient for patients. He thoughtfully included a drain so that invalid patients would not be drowned if a leak developed in the bladder.

In the drawings, Figure 5 shows layers within the bed with a pump for filling or draining on the right (shown in more detail in Figure 6). In the bed, 53 represents the liquid, 46 being the thermostat. Figure 7 is for a heating element. This is held in place by a layer of fibreglass which also insulates the element from any leak. The heat rises up through the liquid. Hall seems to have been unfortunate in exploiting his idea. An appeal case in 1996, Hall v Aqua Queen Mfg., involves a failure to charge seven out of eight infringers of his patent within six years of the alleged infringement.

It has been suggested that the first waterbed was in fact by Neil Arnott, a Scottish surgeon who designed one that was used in a London hospital in 1873. The idea was to prevent pressure ulcers (bed sores). This unpatented invention was unsuccessful as a suitable material for holding the liquid in place had not yet been designed. An obvious problem is in the seams in the cover, which are liable to suffer from fatigue in use as they flex repeatedly. It is suggested that seams should be placed as far as possible away from the areas of use. An interesting variation on floatation tanks is US 4266101, published in 1987, which is called a waterbed womb. The user slides into it and is encased in water while staying dry.

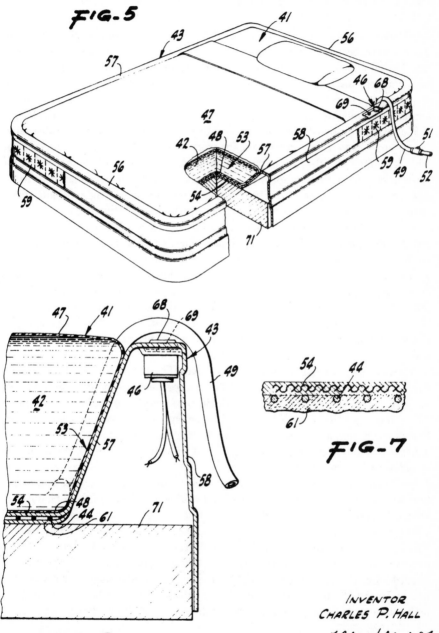

FIG.-5

FIG.-6

FIG-7

INVENTOR
CHARLES P. HALL

BY Flehr, Hohbach, Vest,
Albritton & Herbert
ATTORNEYS

The Workmate® workbench

Versatile and portable workbench
Ronald Price Hickman, Waltham Abbey, Essex, England
Filed 4 March 1968 and published as GB 1267032–5 *and* US 3615087

Ron Hickman was a South African who, although without formal engineering or design training, became head of design for Lotus Cars, which made expensive sports
cars. He left the company when he wanted to explore ideas of his own. He was a do-
it-yourself enthusiast, and one day he cut a chair in half while using it to saw a piece
of plywood. He worked on the idea of a versatile workbench which would be
portable and foldable, so that you could take it where you needed it, and easily store
it. He worked on several prototypes, and by chance had two narrow beams rather
than a single large beam to hand for the work surface when he made the first one.
This developed into the idea of parallel beams forming the work surface where one
beam moves at an angle to form the vice, rather than the usual attachment to one
side. He credited the development and features to his not being a designer or inventor of workbenches, so that he could use fresh, 'lateral' thinking.

He offered the idea to eight companies but all rejected it. Stanley Tools' reply
was that the potential 'could be measured in dozens rather than hundreds' of units.
He began manufacture himself in an ex-brewery, and in a few years was making
14,000 annually. In 1972 Black and Decker, the power tools company, approached
him. They had been one of the companies that had rejected it. They asked for a licence. Broadly speaking, Hickman received 3% of the sales price as royalties, part
being copyright and know-how and part patent royalties. It still took a great deal of
effort to persuade the parent company in the United States that the concept was viable.

About 15 incremental patents were filed over the next decade, including a set of
five which were filed together in 1971. These improvements included putting holes
of varying sizes in the work surface in which sticks could be placed for cutting work.
There was much legal work to stop infringements in at least nine countries. In a
British court case, a book press from the World War I era was found not to invali-
date the vice idea as the judge pointed out that the patents said 'workbench' and that
a book press was not the issue.

Hickman later carried out innovative work on ladders and exercise equipment.
One attempt was a combined step-stool and workbench. None of these patented inventions were successful. Perhaps the most ingenious (and simple) was GB
1304372, where a baby's potty had a plastic sheet extending from it along the floor.
The child's weight when standing up would keep the potty on the floor so that the
contents would not be spilled as the wet bottom came away from the potty. It
worked—but the advertising for it did not spell out the advantage, and it was a failure. Over 55 million units of the Workmate® workbench have been sold and Hickman is now a tax exile in Jersey, studying the history of the Hickman family.

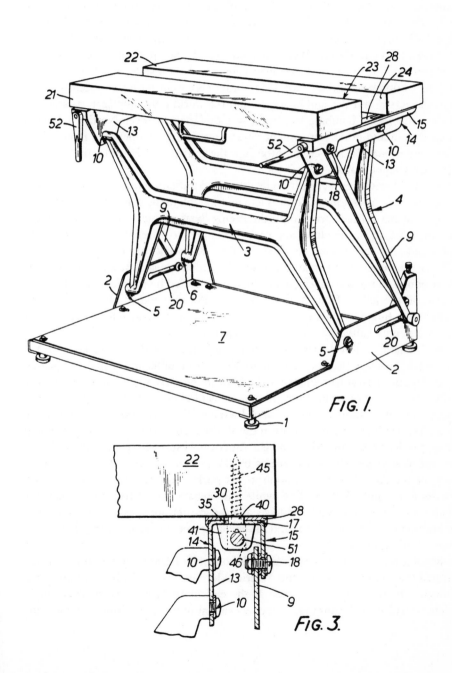

FIG. 1.

FIG. 3.

1970–1979

IF the 1960s began with optimism there was more pessimism in the 1970s and many fierce local wars. The long prosperity enjoyed by Western industrialised countries waned and inflation and unemployment increased owing to economic and industrial turbulence. This was worsened by the oil price rise caused by reduced oil production of Arab states in the wake of the (third) Arab–Israeli conflict in 1973, the Yom Kippur War. Civil war tore Lebanon apart. The continuing problem since 1948 of the Palestinian people seeking a state, and their refugees seeking a homeland of their own, led to bitter clashes in Jordan and Lebanon and terrorist acts of murder and hijack around the world especially against Israel. In Britain the IRA's bombing campaign was designed to sever Northern Ireland from Britain and other terrorists or violent activists were operating internationally: the German Baader-Meinhof group, the Italian Red Brigade, the Japanese Red Army among them.

The USA withdrew at last from its moral and military morass in Vietnam; but, after American withdrawal, South Vietnam soon fell to the communist North creating a further refugee problem in the fleeing 'boat people'. Armed involvement by the USA in Cambodia (later named Kampuchea) destabilised government there and after the ensuing civil war Pol Pot's Khmer Rouge held power. As the Khmer Rouge sought to create an agricultural communist state and eliminate the more educated town-dwellers, 2.5 million people died by murder, starvation and disease in the killing fields of Cambodia. Pol Pot thus joined Hitler, Stalin and Mao in the foremost rank of ruthless employers of mass-slaughter. Cambodia's agony was only halted by invasion from Vietnam.

A dramatic political crisis engulfed Washington with the scandal of the Watergate burglary that forced Richard Nixon from office in the first resignation of an American president. People in the USA in the early 1970s had seen evidence of massacre by their own troops, photographs of children fleeing napalm bombing, peace without much honour: all in Vietnam. They had also seen wrongdoing at the highest level of government in the White House with both a president and vice-president resigning. But younger Americans especially were reassured during the Watergate televised hearings by the presence of Sam Ervin carrying a copy of the American Constitution in his pocket and quoting from the Bible, Shakespeare and William Pitt while his Senate committee unravelled the Watergate conspiracy. They could see too the greater openness of government in publishing the *Pentagon Papers* and the encouragement of more open relations between China and the USA.

Revolution in Iran in 1979 brought Khomeini's Islamic regime to power and created the circumstances for the American hostage crisis which undermined the influence of the USA world-wide at the end of the 1970s, as involvement in Vietnam had at the beginning. There was violent upheaval in Chile with the murder of its president Allende, revolution in Nicaragua and torture and murder of dissidents and the innocent ('the disappeared') by military units in Argentina and Chile. Uprising against the Kabul government by Islamic rebels drew Soviet forces fatefully into Afghanistan.

Bangladesh became independent from Pakistan and Angola and Mozambique secured their independence from Portugal's last vestige of European colonialism in Africa: all involving civil war, which also flared across Ethiopia. Rhodesia's Smith regime became increasingly embattled in the face of patriotic front opposition and independence, as Zimbabwe, followed in 1980. Rioting swept South Africa, especially in Soweto.

There was progress, however, among the superpowers on limiting and reducing nuclear arms via the SALT treaties and by the Helsinki accords between NATO and the Warsaw Pact. This process was disrupted by the Soviet invasion of Afghanistan. In Europe there were the first direct elections to the European Parliament but economic weaknesses became more apparent, especially in Britain which experienced high inflation and strikes leading up to a 'winter of discontent' in 1979.

Nor was Britain by any means the only country in the West with these problems. The 1970s were to confirm the beginning of much reduction and rationalisation of industry some of it, like shipbuilding and steel, moving increasingly to the cheaper Far East. It was also a transitional decade for the electronic revolution gathering pace, on a journey from 'rustbelt' industry to 'sunrise' ventures. Increasing reminder of the dangers of technology lay in the radioactive leak at Three Mile Island in the USA, escape of poisonous dioxin gas from the Seveso plant in Italy and the pollution disaster on the French coast from the wreck of the oil tanker *Amoco Cadiz*.

Men drove on the Moon's surface, probes landed on Venus and took photographs of Mars and Jupiter, the Skylab space station was launched and the Apollo and Soyuz spacecraft linked up in orbit reflecting greater US-Soviet cooperation in space. The everyday rush of inventions continued: Sony invented the Walkman and the first known video game (*Pong*) was used. IBM made floppy disks and personal computers (the first being the Altair 8800) came on the market, along with spreadsheet programs which were increasingly used by business. Smallpox was announced by the World Health Organisation to be at an end and Legionnaire's disease was identified. Louise Brown became the first test tube baby.

The artificial heart

Pump installed to replace or supplement the heart
Robert Jarvik for the University of Utah, both Salt Lake City, Utah
Filed 9 December 1977 and published as US 4173796

Robert Jarvik is a pioneer in the new technology of artificial hearts. The first human heart transplant occurred in 1967 but there have also been attempts to engineer an artificial heart to replace, or to supplement, the human heart.

Jarvik begins his patent with an extensive review of what had already been done in the field. Over 100 American patents had already been granted on artificial heart pumps, besides attempts to provide a mechanical diaphragm such as US 3896501 with many providing different power sources which he cites. He also mentions the many factors that have to be taken into account in designing an artificial heart, including avoiding rejection by the body, preventing excess heat generation, and keeping the noise level low. His work, which he mentions was carried out in calves, was thought of as a supplement to diseased hearts, perhaps as a preliminary to a transplant, rather than a total replacement of the heart. An artificial heart consists of both the pump itself and a means of generating power, but in this patent Jarvik assumed that there would be an external power source.

Jarvik is unusually detailed in explaining the advantages of his proposed heart. The figures show an axial flow pump with a brushless DC motor in Figure 1 with a schematic diagram of it being incorporated in his system in Figure 2. An axial flow pump is normally used in large-scale models and his small device would need careful engineering but he prefers it to the more normal and easier centrifugal pumps as it allows for a change in flow, as shown in Figure 1. 22 marks the 'impeller' and 24 shows the rotation. By reversing the rotation the flow is reversed. Much detail is given in the patent which it is not possible to discuss here. The heart works by alternating contractions and relaxation. Figures 6 and 7 both show the heart with the device at the bottom, immersed in a 'hydraulic fluid', preferably water, shown as 34 within a sac. The pump mimics the heart with the aid of electronics so that the two figures show the sac expanding and contracting to force the blood in and out.

The Jarvik-7 model was the first to be installed to replace a heart, in 1982. The patient died after 3 months, and the next four patients did not survive long and the experiment was called off. Many patients have experienced strokes. Jarvik is still working in the field and recently launched his Jarvik-2000 model. The size of a thumb, it is connected by a very thin power line to a transmitter the size of a small coin which is screwed into the skull behind the ear. A battery pack is worn around the waist. It costs £50,000 to install one which may sound expensive, but that is cheap when the cost of providing for a patient with heart disease is considered.

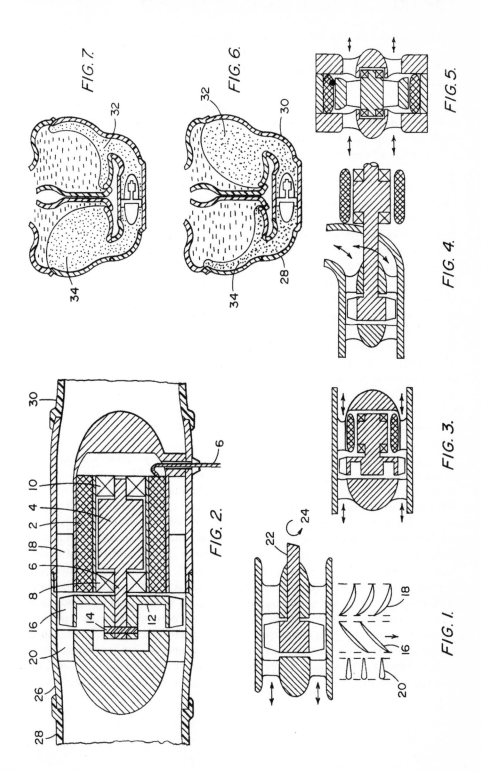

FIG. 7

FIG. 6

FIG. 5.

FIG. 4.

FIG. 3.

FIG. 2.

FIG. I.

The dual cyclone vacuum cleaner

Bagless vacuum cleaner using cyclone technology
James Dyson, Bath, Somerset, England
Filed 14 April 1979 and published as EP 18197 *and* US 4373228

James Dyson was educated at the Royal College of Art as a designer. He moved from designing into engineering concepts while keeping his interest in attractive products. He designed the Ballbarrow, a wheelbarrow with a single huge ball as a wheel, the idea being that it would not get stuck in difficult ground or leave grooves in the ground. It did not sell well as the market is small and wheelbarrows take a long time to wear out. One day he noticed that the air filters in the spray-finishing room for the product was constantly clogging with powder particles. He built an industrial cyclone tower, which removed the particles by centrifugal force, spinning them in air at the speed of sound.

This was not new technology, but Dyson wondered if he could use the same idea for vacuum cleaners, as he was having a similar problem at home while cleaning. Conventional machines have bags which filter the incoming dirty air. They are often awkward and messy to install and remove, frequently have to be replaced, and power is quickly lost when there is still plenty of room to store the dirt. Dyson spent five years and built no fewer than 5,127 prototypes designing a dual cyclone. It is 'dual' because there are two different cyclones, one being for the larger particles. The debris is deposited in a chamber outside the cyclones and can be seen through clear plastic: the reverse of the usual tendency to hide dirt, but it does show that it is working and when it needs to be emptied. There is no loss of power while the debris accumulates; more dirt is removed because of the strength of the process; and dirt-impregnated air is not allowed to circulate as can happen with conventional models.

He approached banks for finance but they were not interested in backing an apparently risky idea. Many manufacturers also rejected him. He applied for patents in many countries and his family went through lean times because of the costs. In order to afford the tooling-up costs of making the product he had to sell the Japanese rights, where it was priced at £1,200 (and sold well). Sales rose and in 1993 a combined factory and research establishment was built at Malmesbury, Wiltshire. Its work ethos is a little unconventional: its engineers are deliberately hired straight from university, staff are encouraged to cycle to work, and suits are discouraged.

Finance was very difficult at first but the result is that Dyson owns the entire company. By 1995 he was outselling the competition in Britain in sales income although the product sells at about £200, significantly more than for other machines. Numerous patents have been taken out for minor improvements and different models are available. There has also been litigation against alleged infringers. Dyson feels that the patent system is too expensive, and that the payment of renewal fees to keep the patents in force is unjust. He is waging a campaign on this subject.

0018197

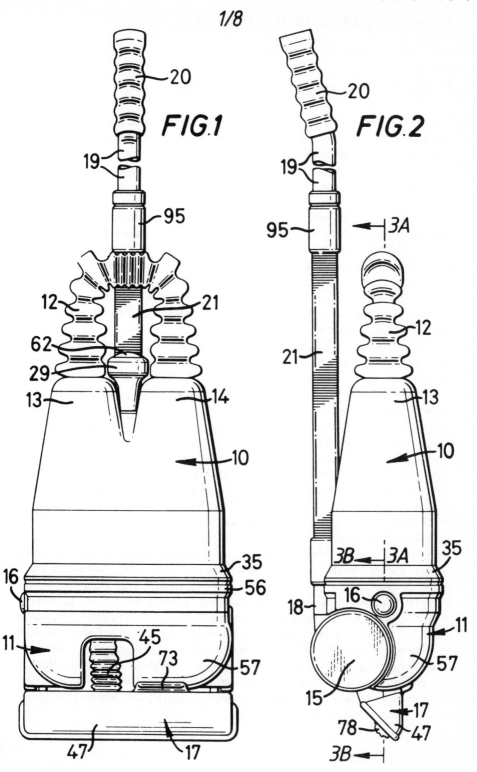

FIG.1

FIG.2

Magnetic resonance imaging

Detecting cancerous tissues in body cells
Raymond Damadian, Forest Hill, New York
Filed 17 March 1972 and published as US 3789832

Computer-aided tomography (see page 150) was a big advance in medicine as it provided three-dimensional pictures of tissues. However, it still relied on X-rays, which are less informative for soft tissue such as cancerous tumours. Its use means that it is often only possible to detect that the patient has cancer, rather than to detect how far it has spread. Raymond Damadian was born in 1936 and trained as a doctor, specialising in cancer. He also had extensive experience in physics. In 1970 he discovered the difference in 'relaxation times' between healthy cells and those that are chemically different such as cancerous cells if they are magnetically stimulated. This involves the nuclei being affected by magnetism (they are in effect tiny magnets) and measuring the time which they take to return to normal, with cancerous cells taking much longer. The principle of nuclear magnetic resonance had been known since the late 1940s and involved studying molecular structure. He invented a way to use that technology to make images of cancerous cells so that appropriate treatment could be devised.

The idea is that the patient is exposed to a static and known magnetic field and at the same time to an oscillating magnetic field. In Figure 2 the patient is exposed to a magnetic field 23 and also to one conducted by transmitter probe 70 which has a beam focusing mechanism 71. The probe gradually moves to the bottom along the track 72 and is always at right angles to the field 23. The nuclei are energised to higher energy states and measurements are taken of the time required for them to return to equilibrium. The areas that take the longest to return are the most malignant. The apparatus on the left analyses the data and presents an image.

Damadian's invention was improved with his US 4354499, which involves the now conventional treatment of the patient lying in a tunnel, and by many later patents. It was only in 1977 with his machine 'Indomitable' (now in the Smithsonian) that the first image of an actual person, his colleague Lawrence Minkoff, was obtained using the new technology. The initial term, nuclear magnetic resonance imaging, was quietly changed to the current magnetic resonance imaging (MRI) when it was realised that patients were getting scared when they heard it, although in fact X-rays have far more impact on the body. In 1978 Damadian founded the FONAR Corporation to market his invention. Times were difficult initially, as other companies were alleged to have infringed the patents. Hitachi settled out of court, but a court case with General Electric began in 1992 and ended in 1997 with the Supreme Court ruling that US 3789832 had been infringed. $128 million was paid out in damages. The technology is now well-established. The pictures that can be derived are 10 to 30 times more detailed than those possible from X-rays, and they are also in colour. They are the attractive pictures you often see in magazine articles.

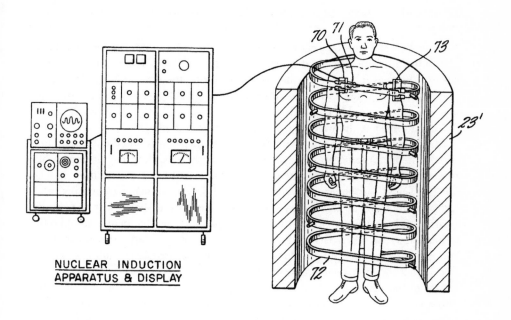

NUCLEAR INDUCTION
APPARATUS & DISPLAY

FIG. 2

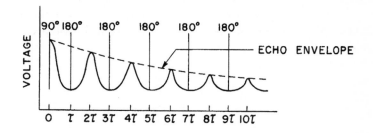

FIG. 3

Personal stereo

Portable recorded music
Andreas Pavel, Rome, Italy
Filed 24 March 1977 and published as **GB 1601447** *and* **US 4412106**

No, this isn't the familiar personal stereo. It is, however, a forerunner, and the applicant has attempted to assert his rights over that of Sony. The famous Walkman® has its origins as follows, according to Akio Morita, the head of Sony. He was sitting in his office when a top aide came in with a portable stereo tape recorder and a pair of headphones. He complained that they were too heavy for what he wanted: to listen to music as he went about without disturbing others.

Morita realised that many young people loved their music and did not want to be without it. He ordered an engineer to strip out the recording circuit and speaker from one of their existing cassette players and to install a stereo amplifier. He asked for lightweight earphones to be added. The advantages of using an existing model were that most of the technology was already known and spare parts were readily available. There were sceptics within the company, saying that people would refuse to buy a cassette player that could not record. Morita pointed out that car stereos also did not have recording ability. He was confident that large volumes could be sold so that the price could come down rapidly.

One of the advantages of a personal stereo is that only a tiny amount of battery power is needed to give sound as there is no need to transmit the sound to an entire room, just to a pair of ears. Morita tried out the first expermental model and, seeing the annoyance on his wife's face, thought that it ought to be modified so that two people could listen at the same time. Later he asked for a button-activated microphone so that two people could talk to each other over the music. Sony's marketing staff were convinced that the product would not sell. The name Walkman® was chosen by staff while Morita was away and initially he did not like its ungrammatical flavour, preferring for example Stow Away and Sound About.

The product was launched in April 1979 and sales began to take off. Sony soon realised that people saw the product as very personal, so most units were made so that only one person could listen to it. Prices, initially over £100, began to fall as volumes rose. Over 100 million units were sold. Rival companies began to enter the market as Sony had not patented the product, assuming that it was unpatentable. Pavel began to take action in 1990 in the British courts and elsewhere. The litigation history is complicated, but ultimately he lost his British patent because of relevant prior art.

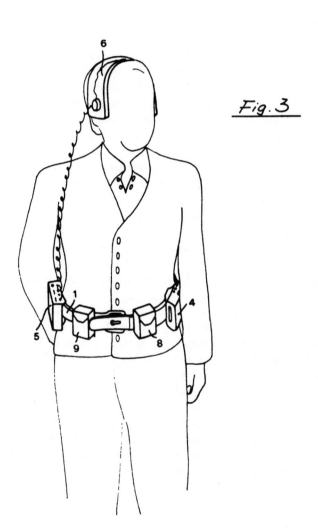

Fig. 3

Post-it® notes

Sticky notepads
Spencer Ferguson Silver, St Paul, Minnesota, for 3M
Filed 9 March 1970 and published as US 3691140

Art Fry was a chemical engineer at 3M, the Minnesota adhesives company. He was in the choir of his local Presbyterian church. He used scraps of paper to mark the selected hymns, but found that he often lost his place because the scraps easily fell out. 'I needed a bookmark that would stay put, yet could easily be removed without damaging my hymnal'.

At this time one of Fry's colleagues, Spencer Silver, was researching a field in adhesives. A mistake was made and the result was a weak adhesive that consisted of numerous tiny particles that was strong enough to hold papers together, yet weak enough to avoid damaging the paper fibres. The patent describes it as 'tacky', and more scientifically as acrylate copolymer microspheres. Provided it didn't get dirty, it was reusable. He asked if anyone could think of a use for a weak adhesive. Fry was shown the adhesive and applied some to the edge of a sheet of paper. He had his bookmark.

Fry happened to write a note on one of his bookmarks, stuck onto a report he was sending to a colleague. The report came back with comments written on the bookmark. He began to promote the product within the company. This was not time wasting, as 3M has a policy allowing its scientists to spend up to 15% of their time on projects they believe in, as this increases work satisfaction.

It was agreed to promote the product in test launches in 1977 in four American cities. The novel packs were sold as at the time the idea of free samples was unheard of. Sales were not good, and as a final effort the 'Boise blitz' was tried out in that Idaho town. People in offices were shown how it worked and given free sample packs. When the salesmen came back a week later nearly 90% said that they wanted to buy packs. The Post-it® was on its way. The product has diversified into 400 different products: 29 colours, 57 shapes and 27 sizes. Yellow is popular as it stands out so well on (white) paper. Some are pre-printed with messages on them.

As it took some time for 3M to realise that they had a winner on their hands, no patent was applied for on the formula outside the United States within the 12-month period required by international law. Although others were free to manufacture it the trade mark was registered in 1976 in the United States and in 1979 in Britain. That initial American registration (there were others) stated that its use was for paper covered on both sides with adhesive which would be stuck on vertical surfaces. Strangely, perhaps, 3M has a non-trade marked product competing with itself.

1

ACRYLATE COPOLYMER MICROSPHERES

BACKGROUND OF THE INVENTION

This invention relates to inherently tacky, elastomeric, solvent-dispersible, solvent-insoluble, acrylate copolymer and a process of preparing the copolymer.

Aerosol spray adhesives have recently found commercial importance in the graphic arts for adhering paper to various substrates, as well as numerous other uses. Such adhesives have many desirable properties. For instance, they permit paper to be removed from a substrate to which it is adhered, without tearing; however, they do not permit rebonding. These adhesives generally comprise solvent dispersions of cross-linked rubbers or acrylates. Such polymers, while commercially utilizable, are not completely satisfactory because the cross-linking reaction is difficult to control and often provides soluble or partially soluble polymers. Soluble polymers are undesirable for spray adhesives having a non-volatile content above 10 percent because they do not atomize well and therefore fail to spray or form a "cobweb" spray pattern. Also, such polymers form agglomerates of random size, the large particles often plugging the spray nozzle orifice. Further, the polymer particles, when dry, agglomerate and are dispersible only with difficulty.

Despite the desirability of inherently tacky, elastomeric polymers which are solvent-dispersible, solvent-insoluble, and of uniformly small size, such a product has never heretofore existed.

SUMMARY

The invention provides inherently tacky, elastomeric, polymers which are uniformly solvent-insoluble, solvent-dispersible, of small size, and ideally suited for use in aerosol spray adhesives. The polymers easily disperse in various solvents to provide non-plugging suspensions which spray without cobwebbing. The polymers permit bonding of paper and other materials to various substrates, permit easy removal of bonded paper from the substrate without tearing, and also permit subsequent rebonding of the paper without application of additional adhesive.

The invention comprises infusible, solvent-dispersible, solvent-insoluble, inherently tacky, elastomeric, acrylate copolymer microspheres consisting essentially of about 90 to about 99.5 percent by weight of at least one alkyl acrylate ester and about 10 to about 0.5 percent by weight of at least one monomer selected from the group consisting of substantially oil-insoluble, water-soluble, ionic monomers and maleic anhydride. Preferably, the microspheres comprise about 95 to about 99 percent by weight acrylate monomer and about 5 to about 1 percent by weight ionic monomer, maleic anhydride, or a mixture thereof. The microspheres are prepared by aqueous suspension polymerization utilizing emulsifier in an amount greater than the critical micelle concentration in the absence of externally added protective colloids or the like.

Solvent suspensions of these microspheres may be sprayed by conventional techniques without cobwebbing or may be incorporated in aerosol containers with suitable propellants such as iso-butane, isobutylene, or the Freons. The tacky microspheres provide a pressure-sensitive adhesive which has a low degree of

2

adhesion permitting separation, repositioning and rebonding of adhered objects. Additionally, these polymers are readily removable from surfaces to which they have been applied, much as rubber cements are removable by mere rubbing. Further, the tacky spheres resist permanent deformation, regaining their spherical shape upon release of pressure. They also exhibit a very low film or tensile strength, less than about 10 psi.

The alkyl acrylate ester monomer portion of the copolymer microspheres may comprise one ester monomer or a mixture of two or more ester monomers. Similarly, the water-soluble, substantially oil-insoluble monomer portion of the copolymer microspheres may comprise maleic anhydride alone, an ionic monomer alone, a mixture of two or more ionic monomers, or a mixture of maleic anhydride with one or more ionic monomers.

The alkyl acrylate ester portion of these microspheres consist of those alkyl acrylate monomers which are oleophilic, water-emulsifiable, of restricted water-solubility, and which, as homopolymers, generally have glass transition temperatures below about −20°C. Alkyl acrylate ester monomers which are suitable for the microspheres of the invention include iso-octyl acrylate, 4-methyl-2-pentyl acrylate, 2-methylbutyl acrylate, sec-butyl acrylate, and the like. Acrylate monomers with glass transition temperatures higher than −20°C. (i.e., tert-butyl acrylate, iso-bornyl acrylate or the like) may be used in conjunction with one of the above described acrylate ester monomers.

The water-soluble ionic monomer portion of these microspheres is comprised of those monomers which are substantially insoluble in oil. By substantially oil-insoluble and water-soluble it is meant that the monomer has a solubility of less than 0.5% by weight and, a distribution ratio at a given temperature (preferably 50°–65C.), of solubility in the oil phase monomer to solubility in the aqueous phase of less than about 0.005, i.e.,

$$D = \frac{\text{Total concentration in organic layer}}{\text{Total concentration in aqueous layer}}$$

Table I illustrates typical distribution ratios (D) for several water-soluble, substantially oil-insoluble ionic monomers.

TABLE I

Oleophilic Monomer	Temp. °C.	Hydrophilic Monomer	D
iso-octyl acrylate	50	1,1-dimethyl-1(2-hydroxypropyl)amine methacrylimide	0.005
do	50	1,1,1-trimethylamine methacrylimide	0.0015
do	65	do	0.003
do	50	N,N-dimethyl-N-(β-methacryloxyethyl) ammonium propionate betaine	<0.002
do	65	do	0.003
do	65	4,4,9-trimethyl-4-azonia-7-oxo-8-oxa-dec-9-ene-1sulfonate	<0.002
do	65	1,1-dimethyl-1(2,3-dihydroxypropyl)amine methacrylimide	0.0015
do	65	sodium acrylate	<0.001
do	65	sodium methacrylate	<0.001
do	65	ammonium acrylate	<0.001
do	65	maleic anhydride	0.02

The recumbent bicycle

Bicycles where the cyclist lies down behind the pedals
Richard Forrestal, Wilmington and David Gordon Wilson, Cambridge for
Fomac Inc., Wilmington, all in Massachusetts
Filed 26 December 1979 and published as **WO 81/01821** *and* **US 4283070**

There has been much research into ways of improving bicycles, the most efficient way of converting human effort into power known. Most of these involve seemingly minor changes, often to benefit a particular type of bicycle, as in shock absorbing for mountain bikes, US 5429344, or handlebars for racers, US 5145094. More rarely efforts have been made to produce a major redesign of the bicycle. One of these was the 'Moulton bicycle', GB 907467, which featured not just smaller wheels (which reduce resistance) but also a major redesign of the way the chassis worked. Harry Bickerton's folding bicycle in GB 1460565 is a graceful attempt to make a bicycle which easily folds up and can be carried around by the handlebars. There are also WO 97/29008, a 'sail and pedal powered vehicle', and US 5342074, a bicycle made for two, but where they ride side by side on a linked chassis.

Another major rethink, going back perhaps to Harold Jarvis' US 690733 in 1901, is an effort to redesign the whole concept by having the cyclist take on a recumbent position rather than sitting erect. Such models are increasingly being seen on the roads. The illustrated patent is an attractive example of the concept, although there are many variations. The reason that the handlebars are missing is that the patent is for the means of adjusting the seat closer or further away to the pedals to allow for people of different heights. The patent gives many reasons why this design is better than the standard bicycle. The main ones are the comfort for the cyclist who has back support on long trips, and safety. The lower centre of gravity and the position of the cyclist mean that in any collision or near collision the cyclist is able to brake more easily, is less likely to be thrown out, can brace more easily with both feet and—should there be a collision—the feet will take the brunt better than say the head or body. Additionally, sharp cornering is easier as the pedals are higher and are less likely to scrape along the ground and (curiously) there is 'the ease by which riders can communicate with automobile drivers'.

There is no claim for higher speeds. Perhaps the greatest drawback is the strangeness of it and the perceived risk of cycling with the head nearer to the ground. There are three main types of recumbent bicycles: those such as this one with a long wheel base, those with a short wheel base, and those with the pedals before the front wheel rather than in the conventional position behind. One of those, US 4659098, is also one of the rarer semi-recumbents, where the front wheel is much smaller than the rear wheel and the cyclist's legs are perched at a 45° angle to the pedals.

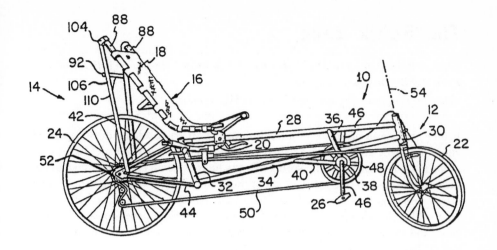

FIG. 1

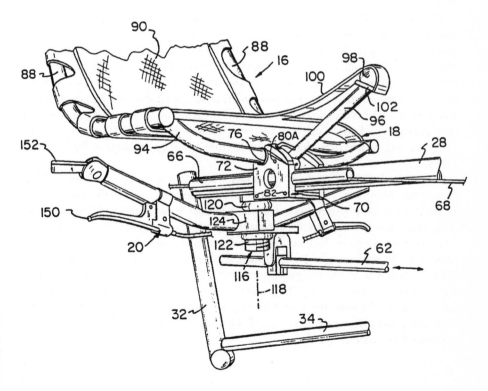

FIG. 2

The 'Rubik' cube

Cube composed of 27 smaller cubes which can be twisted in every direction
Erno Rubik, Budapest, Hungary
Filed 30 January 1975 and published as HU 170062

Erno Rubik was a professor of interior design at an arts academy in Budapest. He enjoyed playing around with geometric cardboard and wooden shapes in his room at his mother's house. In the spring of 1974 he took some blocks of wood and attached them to each other with elastic springs, and started twisting. By the time the elastic broke, he had become fascinated by the changing relationship between the cubes. Then he tried putting differently coloured adhesive paper on each of the six sides, and twisted again. He loved the variety of colours, and then realised he could not get back to the original pattern.

It took a month of intensive work to work out the mathematics and to solve it (clue: he aligned the corners by colour first). He proudly showed it to his mother, who was pleased: now he wouldn't have to work so hard. The finished product involves the cubes being connected by a universal joint-type mechanism. There is only one correct solution and three quintillion incorrect ones. If every person in the world twisted once each second it would take three centuries for a correct solution to arrive by chance.

He took his idea to a small toymaking cooperative in Budapest. Production started on a small scale in Hungary. Then in November 1978 Tibor Laczi, an Hungarian emigré, was shown a cube by a baffled waiter in a café. Laczi bought it off him for a dollar as he liked mathematics. He asked Konsumex, the state trading firm, if he could sell it in the West. They said that there had been no interest in trade fairs. It turned out that they had merely left it on a shelf and not demonstrated it. Laczi went to the Nuremberg trade fair and walked around, twisting the cube and then getting it back to the original colours. Tom Kremer, the British toy expert, was intrigued and helped him secure an order from Ideal Toy Company for a million cubes.

No patent had been applied for abroad within the required 12 months of the Hungarian filing but some protection was available by calling the toy Rubik's Cube®, which was registered as a trade mark in the United States and Britain. However, Ideal Toy got into trouble for patent infringement because Larry Nichols, a Massachusetts chemist, had patented a similar cube (but held together with magnets), US 3655201, in 1972. He had failed to interest toy companies including Ideal Toy. Nichols won an infringement suit in 1984. Over 100 million toys were sold, at least half of them counterfeit. The original company tried to produce them all themselves, but by the time the government gave permission to enlarge premises the craze had passed, and the firm collapsed. To evade the trade mark 'magic cubes' can sometimes be seen for sale.

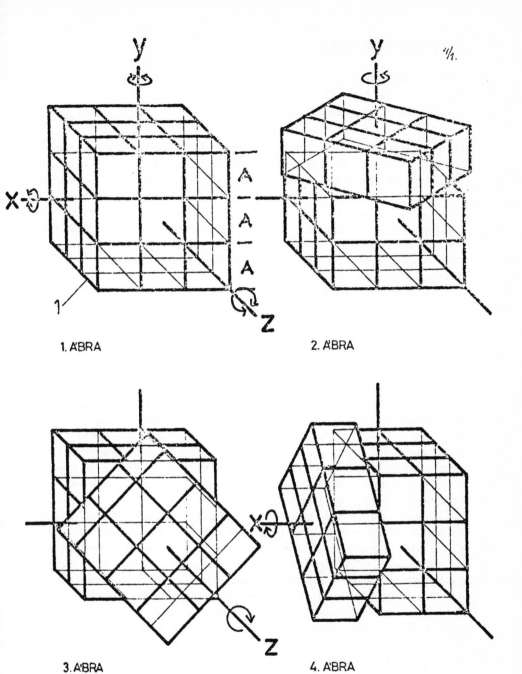

1. ÁBRA

2. ÁBRA

3. ÁBRA

4. ÁBRA

The smart card

Storing computerised data on cards
Roland Moreno, Paris, France for Société Internationale
Filed 25 March 1974 and published as FR 2266222, GB 1504196 *and* US 3971916

Smart cards represent a concept that is much spoken of and yet which is relatively little used at present. Its first practical use dates to anxiety in France over losses from credit card fraud. Counterfeit magnetic strips were being used to make fake cards. Secret codes are on such cards to prevent counterfeiting, but the French were still using old-fashioned 321-bit codes, rather than the much more secure 768-bit codes for validation. Rather than adopt the 768-bit codes, the French government, ten banks and the company Bull CP8 got together to see if a new and more advanced solution could be achieved. The result was the 'carte memoire', or smart card, which was involved in many trials in France.

Roland Moreno was a former science journalist. His invention involves both a portable device such as a card, ring or pen with stored data and a device to exchange data with it, all in a secure environment. He comments on IBM's US 3702464, which, while a smart card, was not designed to prevent illicit reading or unauthorised addition or removal of data, nor to ensure compact storage of data. The illustrated circuit diagram shows to the left of the 'chain line' the storage device, with the circuits needed for the device kept at say a shop on the right. The patent works through in 13 pages of text and nine pages of drawings how the card would act as a kind of debit card. Fresh credit could be added at a shop, and each time a purchase was made the card would be read and its memory altered to show the next shopkeeper how much credit was available. Moreno founded his own firm, Innovatron, in 1974, which is still heavily involved with the technology. Moreno bet 1 million francs that no hacker could 'crack' a smart card. Later improvements were published as US 4007355, 4092524 and 4102493.

The smart card was first marketed in France in 1981 by Bull CP8 and Motorola. By 1993 all French banking cards were smart cards. Smart cards have been slow to gain ground, particularly in the USA. A popular use is opening hotel room doors, as the 'combination' can be readily changed. It is customary now for a single use to be put on a smart card, but there is no reason why many uses would not be possible from a single card, provided that there is cooperation between interested parties and standardisation. It is a nuisance carrying numerous cards around, and in the future people may customise what they want on their own card from a portfolio of applications. These could include a driving licence, finance, medical information and membership details of organisations. A computer terminal or hand-held device would 'read' the required details when necessary. Telephones could be modified to 'read' cards so that they could be charged up with money after contacting a bank. Governments might see the most fruitful use being personal identification and paying out responsibilities such as welfare payments. Civil liberties groups would almost certainly oppose personal identification data being made compulsory.

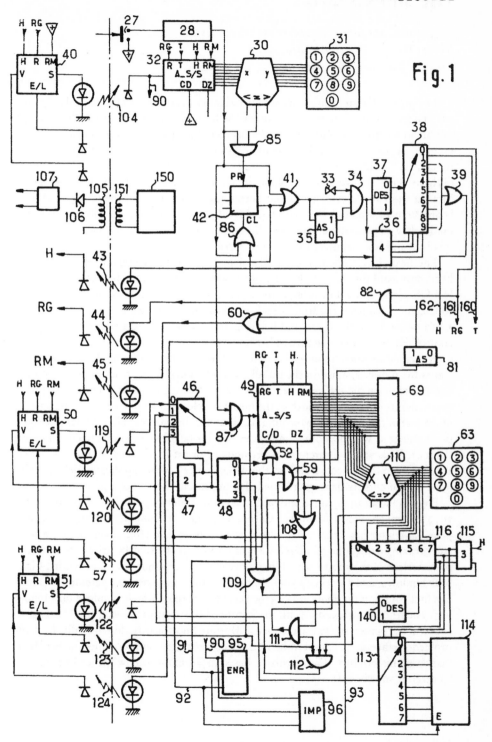

Fig.1

The snowboard

Single, wide ski for use by skiers without poles
Robert Weber, Hyattsville, Maryland
Filed 25 June 1973 and published as US 3900204

The origins of the snowboard are somewhat murky and, though this is probably not the first snowboard, it does appears to be the first snowboard patent which is recognisably as we know it, and it certainly has nice drawings. The stories of the development of the snowboard all involve Americans and all seem to have been influenced by surfing, particularly, as well as by water skiing or skateboarding. None of these sports involve poles to help maintain balanced, unlike downhill skiing. Tom Sims, a 13-year-old school student, surfer and skiier, adapted the idea of a skateboard from plywood to slide on snow in 1963. He later came up with a fibreglass model, like all modern versions. He did not, however, patent it. He continued to be involved and is a manufacturer of snowboards.

The next development was by Sherman Poppen of Muskegon, Michigan. He lived near sand dunes by Lake Michigan, which were covered by snow in winter. His US 3378274, filed in 1966, is for a board of even width down its length with a rope coming up from the front which you were meant to use to hold on. The feet relied on cleats on the board and hanging onto the slender security of the rope for control. He began manufacture of his Snurfer.

Weber's illustrated 'mono-ski' patent is more like the current models and already shows the reduced 'waist' in the middle to aid turning. As with an increasing number of patents on the subject, attention is paid to the way the skier's feet are strapped in. He points out that the width of the board should be at least wider than the length of the feet of the user. In the opening paragraphs of his patent, Weber refers to Poppen's patent as well as another and states that his board increases manoeuvering by the user. This stated awareness of previous prior art is required by the American Patent Office to prevent 'fraud on the patent office', where applicants deliberately leave out potentially damaging knowledge to prevent the loss of a patent.

Poppen began to organise annual Snurfer championships on his sand dunes. In 1979 a stranger called Jake Burton Carpenter turned up from New England to compete. Carpenter had been using a Snurfer at college, and after crashing into a tree and breaking a finger he pondered ways of improving the snowboard. He is credited by historians of the sport with adding more width to the snowboard, providing high back bindings, and a steel edge to help turning, but these were not patented by him. His first patent in the sport is US 5190311, filed in 1991, which is indeed on a binding system for snowboards. With his improved board he easily won the Muskegon contest. Carpenter started a company, Burton Snowboards, but initially he had problems, since as he admitted he initially concentrated on production rather than marketing his product. The company, based in Vermont, is now the leading manufacturer of snowboards.

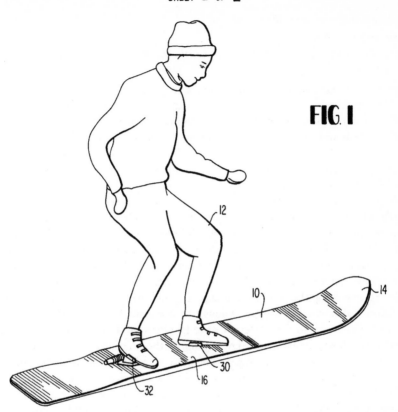

FIG. 1

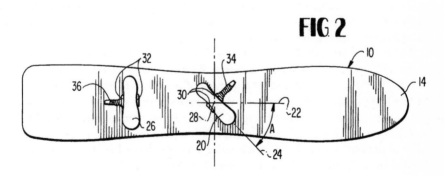

FIG. 2

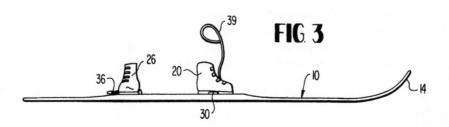

FIG. 3

Wave energy

Extracting energy from waves in the sea
Stephen Salter, Edinburgh, Scotland
Filed 15 November 1973 and published as **GB1482085** *and* **US 3928967**

There are numerous patents concerning extracting energy from the motion of the waves. This particular invention, well-known in energy circles at the time, must serve as an example of a potentially exciting source of renewable energy. The rotation of the earth and the wind give a great deal of energy to the waves. Most attempts to capture some of this energy concern the up and down motion, often with flaps. Others involve the compression of gas by the wave motion, as in Yoshio Masuda's GB 1014196.

Professor Salter led the Wave Energy Group which was set up at the University of Edinburgh in 1974, just after the 1973 energy crisis. He considered these efforts very inefficient in harnessing energy as they did not consider adequately how waves actually behave. He proposed a succession of 'wave engaging members' which would be shaped to remove energy from the wave, and which would pivot about a second mainly stationary member. Figure 1 shows the line of what he called 'ducks' with Figure 2 showing the exterior of one of the ducks facing onto the waves. Each duck, Salter states, should be small enough to use a wave, but the line of ducks should be long enough so that different wave lengths would cancel out. The line is supported by a floating platform. Figure 3 shows the interior of one of the ducks which works within the outer casing, 3. The basic idea is that the oscillation of the duck is converted into mechanical energy by the use of a circular water pump outside 1 in the figure. This would act by 'ribs' such as 9 and 10 exerting pressure on the water between them by using one-way valves. The energy could be converted to electricity by using a turbine mounted on the platform, or transmitted to a station on shore. Salter suggested that deep water was best for his device, as waves lose energy as they approach the shallows.

There was a great deal of interest in this concept. A proposed modification was Salter's US 4134023, for the Department of Energy, where electronics was used to control the duck to help it make the best use of the waves to extract energy. In 1976 Britain's Department of Energy said that it believed that wave power was the most promising of all the renewable energies. In 1982 a consultant reported that the duck, with further development, could be expected to produce electricity at costs comparable to nuclear power. After much indecision the Wave Energy Group was disbanded in 1987 amidst claims by some that the 'Salter Duck' had been treated unfairly in cost analyses by nuclear energy advocates. Professor Salter said in a memorandum to the House of Lords committee on renewable energy, 'We must not waste another fifteen years and dissipate the high motivation of another generation of young engineers'. Unless and until the price of generating electricity again rises steeply, investment in research into renewable energy sources such as wave power is likely to continue to be much less than into other sources.

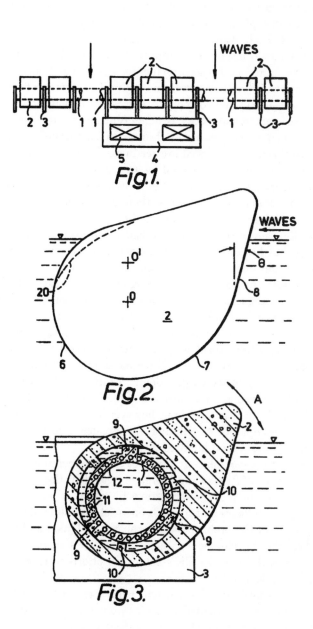

Fig.1.

Fig.2.

Fig.3.

1980–1989

THE Soviet Union's hold over Eastern Europe first relaxed and then disintegrated, and calls for independence were also raised in its own republics. In Armenia and Azerbaijan there was an explosion of ethnic conflict. Poland's free trade union alliance, Solidarity, with Lech Walesa as its leader, lit a beacon for freedom from communist rule that was to inspire many others. Mikhail Gorbachev, the Soviet Union's new premier, introduced reconstruction (*perestroika*) and openness (*glasnost*) to galvanise the rigid Soviet economic and political structures which had led to food shortages, bureaucracy and corruption. The Soviet Union and its satellites battled to balance the weapons and space races, aggravated by the arms build-up between East and West in the early 1980s, with the need to remodel industry and agriculture at home. Gorbachev encouraged greater human rights and the release of dissidents and recognised the economic need for arms reduction.

Ronald Reagan's aggressive stance in the USA towards the Soviet 'evil empire', his 'Star Wars' proposal of a shield of laser weapons around the USA and the deployment of Cruise mobile nuclear missiles in Europe reinvigorated the peace movement. Crucially, they also disturbed the balance of nuclear deterrence between superpowers thus further pressurising Soviet resources. A series of superpower summits brought increasing agreement between East and West and significant arms control of nuclear weapons through the INF Treaty. All this, plus Soviet withdrawal from the military quicksands of Afghanistan, opened the way for the Soviet Union to deal with its internal problems without fear of attack from the West. But once reform had begun, Eastern Europe's communist rulers were 'carried away by the gale of the world' as governments and constitutions changed rapidly and in general peacefully. Romania's regime sought to resist the advance of the new democracy only to see its president, Ceauflescu, deposed and shot. The year 1989 may be viewed as the formal end of the Cold War with the tearing down of the Berlin Wall. In that same year, however, China's troops fired upon students and young people demonstrating for greater reform in Beijing's Tiananmen Square, killing several hundred people.

These immense changes took place at a time of close cooperation between the US and UK governments of Reagan and Margaret Thatcher. Both wanted to revitalise their countries' economies and reputations by reform of 'big government', labour relations, ageing industry and through financial discipline. They came to understand, too, many of the problems facing the Soviet Union and to support broadly the far-reaching developments which Gorbachev's vision set in train. Thatcher gained status through the UK's successful efforts to expel Argentina from the Falkland Islands in 1982.

The Middle East remained explosive. Anarchy held sway in Beirut as both Israel and Syria entered Lebanon, and many Palestinians were murdered in their refugee camps. A land dispute gave Saddam Hussein's aggressive regime in Iraq an excuse to invade Iran and launch a bloody war of attrition that also threatened international oil shipping. With both countries exhausted, a ceasefire was negotiated in 1988. Other Arab states had supported Iraq in their alarm at the possible spread of resurgent and militant Islam which had inspired Iran's resistance and reinforced its unity.

South Africa, the last stronghold of white rule in Africa, began to relax its racial laws as economic sanctions from around the world isolated its government further. Long-lived dictatorships were overthrown in Haiti, Paraguay and the Philippines while Argentina ended military rule in the wake of its disastrous Falklands war. Violence and terrorism, however, continued to threaten world leaders. Sadat in Egypt and Palme in Sweden were murdered and there were attempted assassinations of the Pope, Reagan and (by the IRA) Thatcher.

On Black Monday (19 October) 1987 the long rising market in share values ended and 600 million shares were traded in the race to sell stock. A one-day fall in values in the USA was twice that of the crash of 1929. In a sign of the more robust and organised economic and banking structures of the developed world not only did a world crash not follow but many businesses and stocks were to recover strong levels before long.

The world was threatened more by environmental than by economic disaster. The accident in 1986 at the Chernobyl nuclear power station in the Soviet Union contaminated large areas of Europe through its wind-assisted radioactive fall-out. High radiation levels were measured in Sweden and over a decade later there were still cases of lambs born deformed in North Wales in the UK. Damage from industrial and chemical activity to the Earth's ozone layer was calculated as worse than previously thought and there was evidence of an increase in human skin cancer. The increasing destruction of tropical rain forests such as the Amazon in Brazil caused great concern because of its consequences for the 'greenhouse effect' which could introduce unpredictable and dangerous changes in climate. Further man-made disasters included the pollution caused when the oil tanker *Exxon Valdez* ran aground off the US coast and a serious chemical spill into the river Rhine from an industrial plant in Basle in Switzerland.

Nevertheless, great technical and medical advances continued. The high speed (TGV) train was introduced in France and a microlite aircraft circled the world non-stop. The first public international electronic facsimile service was launched (in a technology that could trace its origin back to the 1840s) and a Cellphone network introduced in Chicago. Microsoft developed a disc operating system to use with IBM personal computers, to become an industry standard, and the first glass-fibre optic cable was laid across the Atlantic. Music compact discs went on sale, Harvard University was granted a patent for a mouse developed by genetic engineering and the first saleable genetically modified product, insulin produced by bacteria, was marketed by Eli Lilley. Astronomers discovered the most massive star yet traced, R136a (2,500 times larger than the Sun), and the Voyager 2 spacecraft sent back pictures of the planets Uranus and Neptune. Man's ingenuity in pursuit of good or ill was revealed again in the first viruses planted by 'hackers' within information technology systems.

Anti-theft devices for cars

Devices for immobilising vehicles to prevent theft
Mosheh Tamir, Tel Aviv, Israel
Filed 27 November 1985 and published as US 4699238

No book on modern inventions would be complete without an anti-theft device for cars. By necessity this is a growth industry for patenting, and this is merely one of thousands of such inventions. What exactly is immobilised to prevent theft will vary. The famous 'Denver boot' or wheel clamp (US 2960857) was originally thought of as a way to prevent the theft of cars and it was only later that it was thought of as a way of preventing the owner from driving away. A simple precaution is having superior door locks. For a long time these were not used by most car manufacturers even though the additional cost was minimal, and they deter many opportunist thieves. Probably the most usual method is immobilising the steering wheel, as illustrated in this patent. Tamir points out that many such devices have hooks round the steering wheel and can be removed by cutting through the steering wheel rim. His invention has the top portion round the central hub with the bottom end wrapping round the brakes 128 and the clutch 130, the cutting through of which would have drastic consequences.

Anti-theft devices can be divided into those which have to be activated before leaving the car, such as steering wheel or column locks, and those which have to be deactivated on the return. In both cases the driver has to remember to do something. A few are in neither category. Examples of different types of preventing theft are GB 2023897, where a code must be entered; WO 89/0518, where a spray is fired at the thief's eyes; US 4969342, where the tyres are deflated; US 5574424, where a driver of unauthorised weight is detected when sitting down; EP 559054, which leaves the driving seat locked forward so that nobody can get into the seat; GB 2233487, where the fuel supply is cut off; US 54866806, where the speed is kept low if a code is not entered; GB 2008192, where the brakes are immobilised so that they are kept on when first applied by a thief; US 4462859, where access to the ignition lock on the steering column is prevented; and GB 2185456, where access is denied to the pedals. If you think that many of these methods are unfamiliar to you it is quite possible that the invention never went into production.

Another possibility is removing a vital piece of equipment. This would be preferably attached to the ignition key to prevent the driver having to remember to take something when leaving. Another division is between those devices which are clearly in view and those which are hidden from view. The latter only work when the car has actually been broken into, and will hopefully confuse the thief enough to cause him to abandon the car, having not spotted the device so as to try to defeat it. The former may act as a deterrent as they are obvious. Fresh ideas are continually being thought of, such as scanning the face of a driver and only permitting use by authorised drivers, a system which can allow for sunglasses and hats, and the famous invention where the thief is locked in, perhaps to cause much more damage while he awaits the police—if he doesn't manage to break out in time.

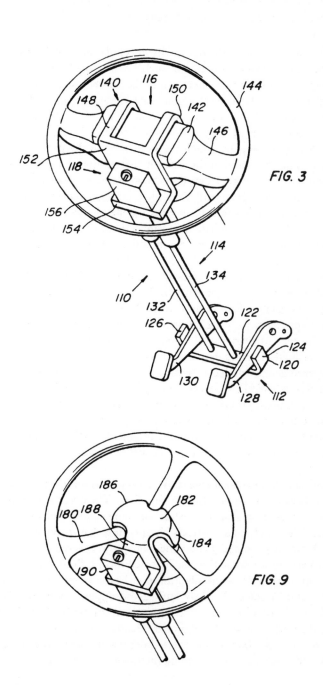

FIG. 3

FIG. 9

Cellular phones

Telephones operating without landlines
Jouko Tattari for Nokia Mobile Phones Ltd, both of Salo, Finland
Filed 25 May 1989 and published as US 5265158

There has been an explosion in the popularity of cellular or 'mobile' phones, which are now *the* fashion accessory. Originally such phones involved making a call by radio to an operator who had to connect you to the number required. Nowadays by relying on networks of 'cells' of space around each aerial structure calls can be made from or to most populated parts of the industrialised world. Such cells are normally pentagons but WO 98/53618 suggests six-sided cells. A leading company in the technology is Nokia, a company which began in 1865 in the paper industry. The company began to diversify and in the 1960s it entered the telecommunications industry. Finland now has a higher density of mobile phones than anywhere else on Earth. The drawing shown is an early example of a 'stand alone portable telephone unit'. The patent actually has to explain what mobile phones are while showing in Figure 1 a phone which is still reliant on a connection to a power supply and a link to telecommunications, and in Figure 2 a modified version with antenna and batteries.

The first generation of cellular phones appeared in 1979 in Japan. There were no standards and technology was primitive (and expensive) by today's standards. In 1982 and 1983 they came to Europe and North America. In the early 1990s the second generation appeared, with digital instead of analogue phones and a common standard for Europe. Analogue is like voices while digital is the way computers function, and facilitates moving data around as well as making eavesdropping more difficult. The third generation has recently started: communication with the Internet is possible through WAP, Wireless Application Protocol and the related WML, Wireless Markup Language. Geoworks claims to have a monopoly on such technology with its US 5327529 which enables these two passions to be combined. Incidentally, Tim Berners-Lee, who wrote the original HTML language used on the Internet, deliberately did not patent it to encourage its spread (otherwise it would be in this book).

In Britain at least the cellular phone scene is very confused with literally millions of possible tariffs available from different suppliers. The quick spread of pre-paid phones rather than tariffs has made the situation more comprehensible. British mobile phones are deliberately sold relatively cheaply (less than £100) although the actual costs are between £200 and £500. The money is recovered by charging much more for the calls than would be the case in, say, Finland. Pre-paid phones are said to be popular with criminals as no identification has to be produced and calls are difficult to monitor or trace. An early anti-eavesdropping phone is GB 2021355 while WO 98/49855 and US 5722067 require authentication by the user to prevent thieves using the phones, a sad development. Hands-free mobiles are now becoming popular, as with US 5841856, and the sight of people wandering around having conversations with themselves is becoming increasingly common.

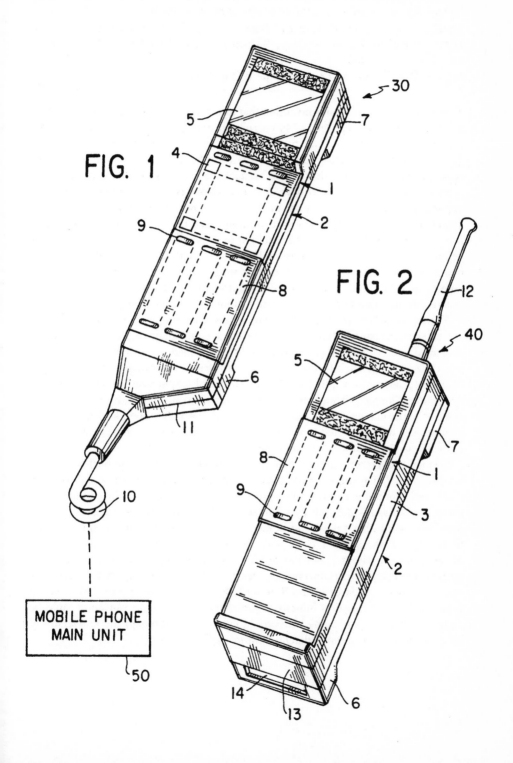

FIG. 1

FIG. 2

MOBILE PHONE
MAIN UNIT

Cold fusion

Generating power from combining atoms at room temperature
Stanley Pons, Cheves Walling and John Simons, all of Salt Lake City, Utah
and Martin Fleischmann, Tisbury, Wiltshire, England for University of Utah,
Salt Lake City, Utah
Filed 13 March 1989 and published as **WO 90/10935**

This is, if you like, the major invention that never was. Its existence was proclaimed in a press announcement on 23 March 1989. It immediately caused a furore, especially as it is normal for scientists to publish their results first in a paper in an academic journal. It was later suggested that the University of Utah asked for the announcement to be brought forward because the secret looked like leaking out.

Pons and Fleischmann were both university chemists, rather than physicists, who might have been expected to work in the field of power generation. Nuclear fission in atomic power plants involves splitting atoms, and the opposite process is fusion. It is what the sun does. It was only thought to be possible at fantastically high temperatures, and much fruitless research had been carried out. A great advantage over fission would be that since fusion reactors would need great efforts to keep a very high temperature, they would shut down automatically if anything went wrong, as such artificially high temperatures would suddenly no longer be generated.

The few details given were tantalising. It was like a simple experiment which involved electrochemical techniques to fuse some components in heavy water which contained deuterium. It was claimed that this happened at room temperature. The press release was more informative about how the idea was worked out on a walk up Millcreek Canyon near Salt Lake City. Pons and Fleischmann worked on the details in Pons' kitchen (it was later said that this was an all-night session over a soon-emptied bottle of bourbon). Later it was revealed that a test–tube–like cell with a pair of electrodes was connected to a battery and immersed in a jar of heavy water (where normal hydrogen isotopes were replaced by heavier ones).

Secrecy was preserved even after the patent application was published in September 1990. Patent specifications have to be detailed enough for those skilled in the same branch of industry to be able to reconstruct the invention to be granted validity later on. It was widely felt by scientists that the more than 100 pages of text and drawings were rather vague, and in fact it was never granted protection. Both the university and the state of Utah were very excited over the possibility of lots of money from the patent. They lobbied for federal funding before any attempt had been made to verify the claims independently. A panel was set up under Nobel laureate and nuclear chemist Glenn Seaborg to look into the claims but they reported negatively. Some scientists initially reported promising results but none could be verified. There are still some believers in cold fusion, although the University of Utah is said to have stopped defending it in 1998.

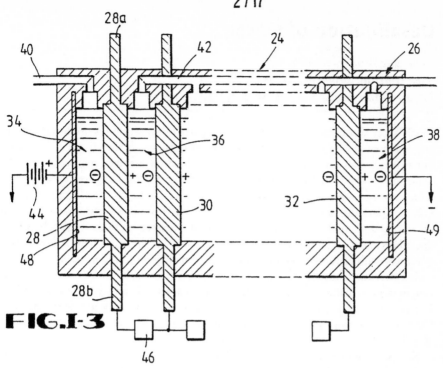

FIG.I-3

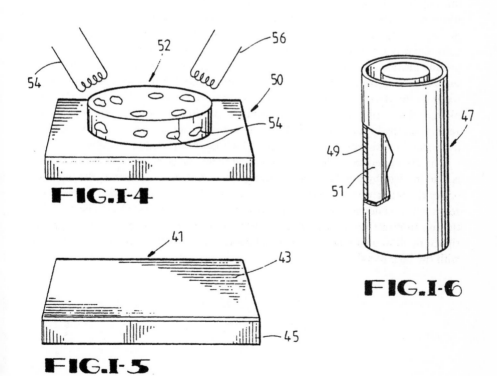

FIG.I-4

FIG.I-5

FIG.I-6

Desalination of water

Removing salts from salt water
Sven Gibelius, Bromma, Sweden
Filed 15 March 1984 and published as WO 85/04159

Fresh water shortages are growing in many parts of the world for use in agriculture and industry as well as in the home. In some of those places there is plenty of water, but it is sea-water, or at least brackish. Desalination involves removing the salts from such water, but has conventionally cost a great deal in energy to carry out. There are two conventional methods of doing so. Multiple-effect evaporators involve evaporating salt-water in numerous long vertical tubes. The vapour is drawn off to cool and condense and the retained heat is passed on to the next tube which does the same with less heat and reduced pressure (as the water boils at lower temperatures with reduced pressure), and then similarly on to the next. As no heat is added after the first tube there are savings in energy costs. The other method, flash-evaporators, is better for larger plants. Heated sea-water is sprayed into a tank which is kept under reduced pressure and hence again evaporation occurs with relatively little heat. A serious problem is a build-up of scale, especially where the temperature is high.

Smaller-scale devices used in hot climates usually involve solar energy. Typically, water within dark plastic sheets evaporates and condenses on the upper sheeting and flows down to be collected in troughs. Other methods, normally in brackish water where there is less salt, involve using membranes which fresh water passes through, using reverse-osmosis as in EP 82705 or electro-dialysis as in US 4776171. Less usual is US 5160634 which uses lasers to vibrate the ions. There are many patents for floating solar stills, as in US 4959127, for collecting water or for use at sea, perhaps in emergency situations. The differences in temperature between the water in the device and that of the surrounding sea-water or the air are often exploited. Solar stills are little use in cloudy conditions or at night, and methods of amplifying the heat are needed. There have been other suggestions such as collecting water from morning mists on nets, which has been tried in Namibia.

The illustrated example is merely one of many suggestions. The inflated tank 1 is held in place by weight 4 with a sheet 3 between them. The heat on the top evaporates the water which passes through 7 to tank 8 where, surrounded by the cooler sea-water, it condenses and passes to tank 19. Gibelius suggests lenses be placed on the top to direct the heat more powerfully onto the tank. The air inlets 5 are designed to allow air to pass through and pass over the water to increase evaporation with the help of valves.

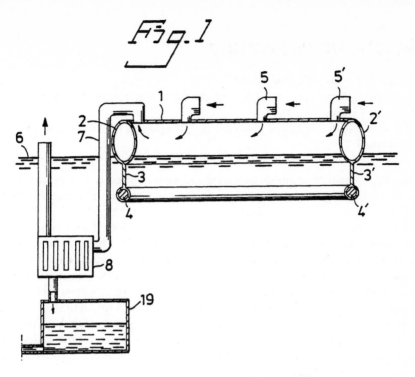

Fig. 1

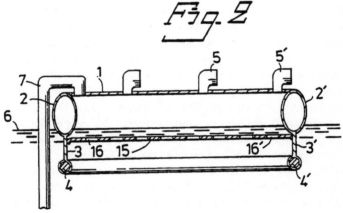

Fig. 2

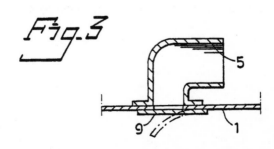

Fig. 3

Genetic fingerprinting

Using DNA to identify individuals
Alec John Jeffreys, Leicester, England, for Lister Institute of Preventive Medicine, London, England
Filed 12 November 1984 and published as **GB 2166445, EP 186271** *and* **US 5413908**

The identification of criminals from evidence such as blood, semen or skin left at the scene had long been an ambition of police laboratories. The fact that DNA within cells provides this was accidentally discovered by Alec Jeffreys.

It had been noticed that DNA had certain sequences which did not contribute to the function of a gene but which were repeated within it. These are called 'minisatellites'. Each organism has a unique pattern except for identical twins. The technique of genetic fingerprinting is a complex one. First the DNA is cut into specific portions by enzymes. The fragments are placed on a gel which is then subject to an electric current. These double-stranded fragments are then split into single strands and transferred to a nylon sheet. They are subjected to synthetic DNA which binds to the minisatellites. They are X-rayed, and the film is exposed. Dark codes show the minisatellites, rather like bar codes on packaged food. It is the pattern of the darkness which is unique.

The possibility of two individuals matching is regarded as millions to one. In addition, related individuals will show much similarity. The technique is used to trace criminals or to prove family relationships and for medical purposes. Genetic fingerprinting was first used for criminal detection by the Leicestershire Constabulary on 5 January 1987. Two schoolgirls had been raped and strangled in two villages. The killer was thought likely to be a local man, and samples of blood and saliva were requested from over 5,000 men. Only two people refused. One of them, Colin Pitchfork, finding himself under pressure to consent, bribed a workmate to take the test in his name. When the substitute later confessed to the deception, Pitchfork was DNA-tested and charged and received two life sentences.

Subsequently, DNA evidence has been used to convict or identify criminals on the evidence of samples kept from as long ago as 1970. Britain has kept a database of DNA evidence from all convicted criminals since the mid 1990s and there are over 500,000 records so far. All those awaiting trial are routinely checked for a match against the database. If convicted, the profile is added to the database. Otherwise it is destroyed. Criminals responsible for murders and rapes from as far back as 1970 have been traced if the evidence has been kept. The British patent was exclusively licensed to ICI (later its breakaway company Zeneca) in 1986.

(12) UK Patent Application (19) GB (11) 2 166 445 A

(43) Application published 8 May 1986

(21) Application No 8525252

(22) Date of filing 14 Oct 1985

(30) Priority data

(31) 8428491   (32) 12 Nov 1984   (33) GB
     8505744   6 Mar 1985
     8518755   24 Jul 1985
     8522135   6 Sep 1985

(71) Applicant
Lister Institute of Preventive Medicine (United Kingdom),
Royal National Orthopaedic Hospital, Brockley Hill,
Stanmore, Middlesex HA7 4LP

(72) Inventor
Alec John Jeffreys

(74) Agent and/or Address for Service
R. G. C. Jenkins & Co., 12-15 Fetter Lane, London EC4A 1PL

(51) INT CL⁴
C12N 15/00 // C12Q 1/68

(52) Domestic classification
C3H B2
U1S 1334 1337 C3H

(56) Documents cited
None

(58) Field of search
C3H
Selected US specifications from IPC sub-classes C12N
C12Q

(54) **Polynucleotide probes**

(57) The invention provides for improved identification of individuals, species etc. by making use of the existence of DNA regions of hypervariability, otherwise called minisatellite regions in which the DNA contains tandem repeat of quasi-block copolymer sequences. The number of repeats or copolymer units varies considerably from one individual to another. Many such regions can be probed simultaneously in such a way as to display this variability using a DNA or other polynucleotide probe of which the essential constituent is a short core sequence, 6 to 16 nucleotides long, tandemly repeated at least 3 and preferably at least 10 times. The probing reveals differences in genomic DNA at multiple highly-polymorphic minisatellite regions to produce an individual-specific DNA "fingerprint" of general use for genetic identification purposes, paternity and maternity testing, forensic medicine and the diagnosis of genetic diseases and cancer.

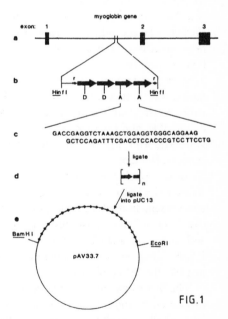

FIG.1

The drawings originally filed were informal and the print here reproduced is taken from a later filed formal copy.

The HOTOL space plane

Air-breathing/ hydrogen fuelled craft for space flight
Alan Bond, Stanford in the Vale, Oxfordshire, England for Rolls-Royce
Filed 23 December 1983 and published as **GB 2240815, 2241319,**
US 5097896 *and* 5101622

Cheap space flight has long been a dream. This is one of the attempts. It was filed in 1983 and ought under British law to have been published 18 months later as an application. However, it was kept secret by the Ministry of Defence until it was allowed to be published in August 1991, apparently after it was pointed out that other countries were working along similar lines.

Alan Bond had been involved in the Blue Streak missile project. HOTOL is supposed to have stood for Horizontal Take-off and Landing but it was modified along the way. His idea involved an unmanned, reusable rocket with the usual nozzles at the rear, but also with small wings. It would take off from a runway with the aid of a rocket sled (which would parachute down to earth when out of fuel). Then until the rocket reached Mach 5 the craft would be propelled by an air-breathing engine where the liquid hydrogen fuel is used to cool the intake of air by acting as a heat-sink. This is made possible by isolating and liquefying the oxygen. The compressed oxygen is burnt to provide propulsion. By now there would be little atmosphere left and the craft switches to using the hydrogen, oxidised by liquid oxygen, to fuel it further. The second patent application by Bond and Bryan Belcher, not illustrated here, describes the heat exchanger. The same pump and nozzles are used in both types of propulsion.

Such an engine is relatively cheap. Hydrogen is a plentiful fuel and the design means that no fuel needs to be stored for the first stage, and there is a minimum of parts. As the craft is unmanned there is no need for a life-support system. It would be reusable, and the plan was that it would be able to take 7 tonnes of cargo into a low earth orbit, just outside the atmosphere. It clearly draws on the Space Shuttle concept. Bond originally thought of launching it vertically from the ground. British Aerospace worked with him on the project. After trying to interest other member states in the European Space Agency in the concept, during 1986 to 1988 the British government studied the concept to see if public funding were worthwhile. They decided not to invest in the project. Collaboration was carried out with the Russians and in 1990 a small HOTOL using known Russian engines was launched in the USSR. This involved its being first carried on the back of an Antonov 225 aircraft, just like the Space Shuttle.

The British patent was granted a few months after publication but it lapsed from protection in 1995. Alan Bond is now the managing director of Reaction Engines Ltd, which is working on an adaptation of the craft called Skylon with partners. This particular project is considered secret and few details are available. A drawing suggests that the engines would be mounted on each wing-tip. Rival projects include India's Hyperplane and McDonnell Douglas and Boeing in the USA.

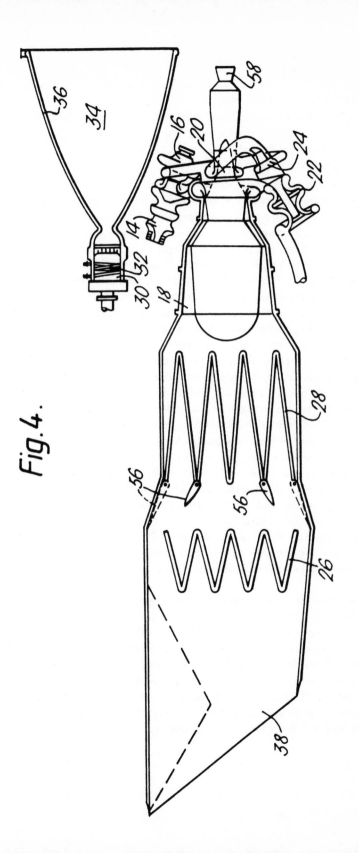

Fig.4.

Solar energy

Using sunshine to produce power
Thomas Burns and Cynthia Burns, both of Fox Point, for Burns-Milwaukee
Inc., Milwaukee, all in Wisconsin
Filed 11 September 1987 and published as US 4848320

It has long been a dream to use sunshine to generate power. The fact that countries in hot climates are often poor, and need cheap power, is an additional incentive. Unfortunately solar power has had many problems fulfilling its destiny, owing to the fact that sunshine is of low intensity across the world's surface and that the cost of providing ways of using it is high. It is expensive to provide ways for a device to track the Sun during the day, for example. This is apart from the problem of no output during the night. Solar cells convert the sunshine into electricity. At first it was thought that the cells themselves needed to be very precisely and formally made until an accidentally distorted set of cells produced much more power. The Sun's rays were bouncing around before being reflected away. US 5833176 is for their use in outer space, where no night-time causes problems. A more common use is for illumination at bus shelters and other places where there is no electrical supply.

A solar collector involves using blackened metal with water or some other liquid running beneath to collect the heat. Half of Israeli homes now have such devices to provide hot water. US 5586548 is for a collector that floats on a swimming pool. Trombe-walls are a variation, as in FR 2578312, where huge barrels of water form all or part of a wall of the house and gather heat during the day. At dusk a panel closes off the exterior, so that heat loss is directed within the house. The same idea for roofs has also been tried out. Trombe-walls are good for places where it is hot during the day but cold at night, but rather spoil the view. Besides such 'active' methods of gathering solar energy, passive designs involve large windows facing south and relying on say tiled floors within conservatories to gather heat. Such materials are 'massive', which means that they slowly absorb and retain heat, unlike wood, which is the opposite. In both active and passive systems fans can be used to blow excess heat down to rock-filled basements from where heat can rise during the night.

An invention for a solar oven is illustrated opposite. The side panels reflect heat into the heating chamber which has a smoked glass door. Critics say that cooking in this fashion takes twice as long as with conventional ovens, and that it must not cloud over, nor should you want to cook at night. They are excellent in deserts where there is no electricity and where firewood is in short supply. Reflecting devices to increase the Sun's intensity can be modest as in this case, or larger such as in US 5460163 which has a trough-shaped mirror with a steam generation tube at the bottom, or huge such as at Mont Louis in France, where numerous mirrors track the Sun and direct it to a central furnace.

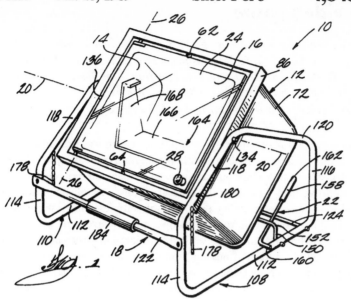

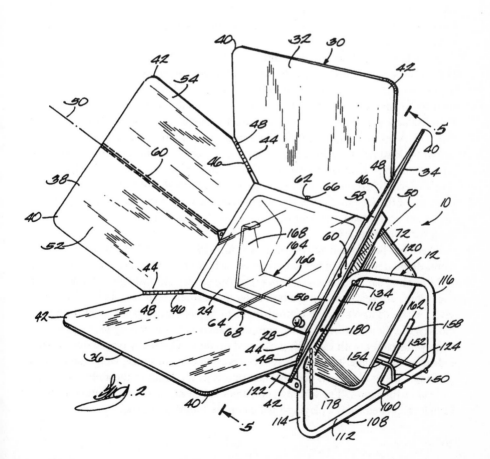

The video game

Electronic game for one or more players displayed on a screen
Roger Hector, Saratoga, Nolan Bushnell, Woodside, Howard Delman, San Jose,
Edward Rotberg, Los Altos and Jon Kinsting, San Jose, all in California, for
Bally Manufacturing Company, Chicago, Illinois
Filed 31 October 1985 and published as US 4720789

The video game was dreamed up by a physicist who was preparing for the annual open day in 1958 at the Brookhaven National Laboratory at Upton, Long Island, in New York state. Willy Higinbotham wanted to make the occasion—which was designed to prove that the staff at the nuclear research establishment did not all glow in the dark—more entertaining than a lot of static exhibits. He thought he could make a simple game out of an oscilloscope, a test instrument which makes lots of wavy lines. He linked it to a black and white television screen and made a bouncing ball appear on the screen. Looking around for spare parts he could use, Higinbotham fashioned a crude game in two weeks of work. Each player would have a box with a knob, and would use it to bounce a ball back and forth over a 'net'. He even managed to get balls hitting the net to bounce back. The result was hundreds of people queuing up to try the game. The next year he changed the game to a larger screen and offered the options of playing on the Moon, with low gravity, or Jupiter, with high gravity. Again it was a great success.

Years later, his children asked him why the family had not become millionaires. He did not patent it because, as he worked for the government, he assumed that he would not get any royalties. The invention of the video game is frequently attributed to Nolan Bushnell, probably for his US 3793483, filed in 1972. He invented Pong (which was similar to Higinbotham's concept but with paddles added) and founded Atari, the first company to exploit the technology. A more interesting patent by Bushnell (for the amateur, anyway) is the splendid example shown here. The patent explains that it provides 'the benefits of healthy exercise'. The weight-sensitive pads correspond to the squares on the screen. The electronics interpret the pressure of the feet and, using the sample game given on the screen, a bug 102, trying to get at the food 100, can be intercepted by the feet 124, which have already dealt with one bug 116.

There has been much activity in this field, with Japanese firms being at the forefront. Rapid development has led to games such as Space Invaders, dating from 1978, and Pac Man, dating from 1983, being regarded as very old-fashioned. Modern games use colour, are very fast and are becoming increasingly realistic—too realistic, some say. The speed at which the controls have to be used are perhaps good training for modern life. Besides feet, buttons, computer keyboards and mostly joysticks are used to play the games, with US 5802462 a good example. Originally played on special arcade machines, sometimes by several players competing in a race, video games are increasingly being played on personal computers as they gain in the ability to run fast and colourful games. The leading companies are Sega, Sony and Nintendo, the last having originated from a playing-card company founded in 1889.

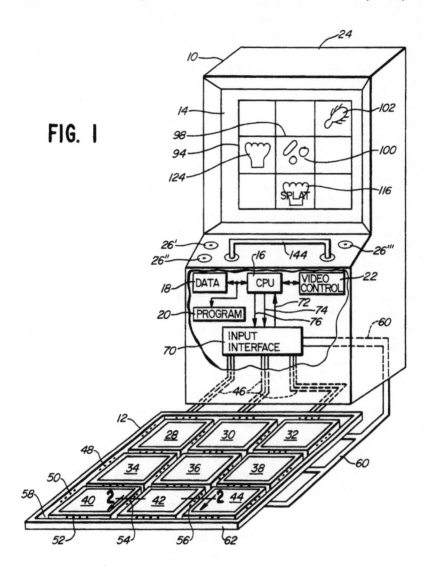

FIG. 1

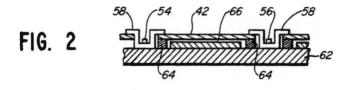

FIG. 2

The 'Widget'

Drink cans which provide a head for beer when poured
Alan James Forage, Beaconsfield, Buckinghamshire, England and William
John Byrne, Dublin, for Arthur Guinness Son & Company, Dublin, Ireland
Filed 29 November 1985 and published as **GB 2183592**, **US 4832968** *and*
EP 227213

Guinness stout is famous for its thick, creamy head at the top of the black, bitter drink, which seems to stay there forever. It consists of millions of tiny bubbles. It has been made much of in its advertising. The trouble was that the head only appeared in public bars and that cans or bottles of the drink could not reproduce the head. This hurt Guinness' sales. The head comes from 'sparklers' in the bar dispensers which add a tiny amount of nitrogen which froths the stout. The nitrogen is stable and does not easily disperse as it moves up through the drink, maintaining the head.

Judging from the patent literature the company worked on producing the head in cans and bottles from 1959. An early version is in GB 1266351, while GB 1588624 discloses mixing carbon dioxide and nitrogen with the contents by one of three methods. These were subjecting the beverage to ultrasonic excitement; injecting gas with a syringe; or pouring the drink over polystryene granules. All resulted in a nice head, but GB 2183592 admitted that all had the defect of extra equipment being needed, and 'It is unreasonable to expect a retail customer to have available an ultrasonic generator'. They spent £5 million on research and came up with the idea of a plastic pod with a tiny hole at the bottom of the can.

The pod is placed at the bottom of the can before it is filled with the drink. Air is expelled and immediately before sealing a tiny quantity of nitrogen is added. The act of sealing plus chilling pressurises the contents and forces 1% of the drink into the pod. When the can is opened, the contents reach normal pressure again. The contents of the pod are forced out and rise up through the drink, and allied with the nitrogen create the familiar and long-lasting head. Carbon dioxide is also present but nitrogen bubbles do not break so easily, hence the head.

Even when the idea of the pod had been conceived work was carried out on the shape, weight and density of the pod. Different prototypes were tested for 18 months, and it was worked out that the hole at the bottom of the pod should be 0.61 millimetres in size. It is however important that the can is chilled when opened, otherwise the contents go everywhere. A warning on the can explains this. The contents should also be poured out in one go. Other brewers have come up with their own 'widget', and the idea was introduced to draught beer in 1992, lager in 1994 and cider in 1997.

FIG.1.

FIG.2.

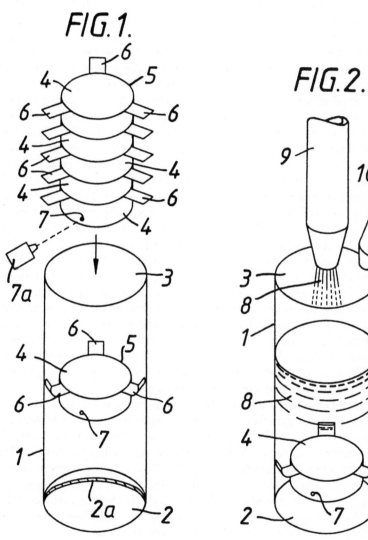

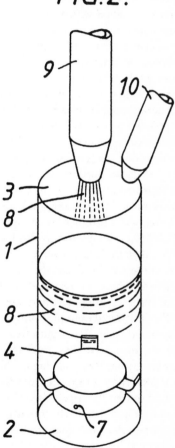

Wind energy

Using wind to produce power
Minoru Abe, Kashiwa, for the Agency of Industrial Science and Technology
and the Ministry of International Trade and Industry, Tokyo, both in Japan
Filed 7 April 1980 and published as JP 56-143369A *and* US 4311434

Wind has been used to produce energy for thousands of years, though it has only been in this century that extensive work has been done on variations on the classic windmill concept. Modern wind-turbines typically fall into two main groups. Three-bladed turbines which face into the wind and are placed to catch the prevailing winds. Two-bladed turbines have the wind behind them and can often swivel to catch the winds. In both cases they are typically situated on coasts or hills in windy areas, and the spinning action turns a shaft which runs a generator. A governor is installed which prevents excessive spinning in high winds, which would cause damage. In a sense a wind-turbine is the opposite of a fan: a fan uses electricity to create wind, rather than using wind to create electricity. The larger wind-turbines can generate up to 750 kilowatts, which is enough to power 300 homes—provided the wind blows. There are other models, such as DE 19719114, where two vertical wind-turbines are connected in the shape of a giant H.

Another popular idea is that of using vertical tubes with air-foils, where reduced air pressure generates power (it is the principle that keeps airplanes flying, after all). They have been suggested for ships as the main means of propulsion as wind from any direction can be used. An example is US 5709419. Less usual are DE 19708624, which suggests putting wind-turbines on the tops of cars and trains to benefit from the slipstream (watch out for bridges), and the spectacular US 5669758 which sites a huge shroud next to the wind-turbine in order to funnel the wind into it.

Often wind-turbines are grouped together, then they are known as wind farms. CH 668623 is an interesting display of numerous wind-turbines on six branches off a central stem. It is of course vital to ensure that each wind-turbine does have access to the wind and that the wind it should receive is not blocked by other wind-turbines. Those in Denmark and California, the most populated regions for wind-turbines, are in fact normally strung out in lines.

The illustrated example is unusual in that the supporting structure is not encased in a smooth skin. Abe argues that even with governors there is a risk of damage in a great storm, as well as problems in maintenance and repair with a structure that may be 50 metres or more high. The nacelle 2 pivots on 8 as the plate 12 slides down on guides 1a by a rope off a windlass 13, powered by an electric motor to near the ground so that the nacelle is vertical with the blades above it for easy care and protection. There has been criticism of wind-turbines, besides the problems which arise when the wind does not blow; conservationists have complained about noise, the harm if blades break off and the visual intrusion on hilltops.

本発明の原理は、風のエネルギーを単に機械的エ
ネルギーに変換する農場型風車装置にも応用する
ことができる。

　以上詳述したように、本発明によれば、タワー
に昇降機構を付設し、この昇降機構における受台
によってタワー上のナセルを昇降させ得るように
構成したので、強風時におけるローターブレード
や回転系の保護及びナセルの保守、点検等を確実
且つ安全に行うことができ、しかもナセルを鉛直
に固定して受台を昇降させるようにしたので、長
大なローターブレードを取付けたままでナセルを
容易に昇降させることができる。

4.　図面の簡単な説明

　第1図は本発明に係る風力原動機の平常運転状
態の斜視図、第2図はナセルを降下させた状態の
斜視図である。

　1 ・・・ タワー、 2 ・・・ ナセル、 3 ，15 ・・・ 座、

　5 ・・・ ローターブレード、　 8 ・・・ 水平軸、

　9 ・・・ 基台、　　　　　　 12 ・・・ 受台。

指定代理人

　　工業技術院機械技術研究所長

　　　　本　田　冨　士　雄

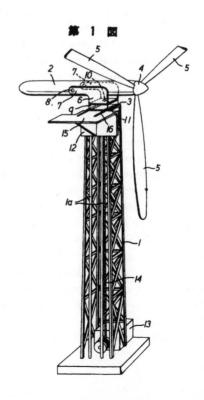

第 1 図

1990–1999

THE close of the 1890s was said among artists and writers to be 'the end of the century, the end of the world'. But the 1990s were to see particular vigour in much of the world's activity: in trade and technical and medical progress as well as heightened optimism for peace and prosperity. This contrasted though with dark patterns of tragedy which continued to disfigure many countries. As communism in Europe collapsed with the Soviet Union ceasing officially to exist in 1991, a new liberty emerged but also old enmity and ethnic tension. Africa and Asia, too, witnessed eruptions of religious and racial killing. The impression of the world both coalescing and fragmenting has rarely been more pronounced. But if many of the hopes of the early decade were tarnished by its end international cooperation and shared concerns, together with the ability to speak freely in so many countries, were still far greater than the years of the Cold War and of Vietnam would have foreseen.

The dismantling of the Soviet Union and consequent independence amongst its 15 republics, due in large measure to Mikhail Gorbachev's policies before 1991 and Boris Yeltsin's reforms after 1993, brought greater freedom to live and to trade and closer links with the West including the START talks limiting the size of the US and Russian arsenals of nuclear weapons. It also led to economic and governmental instability, a growing gap between rich and poor and the end of an expectation of equality which the previous 70 years had fostered. Gorbachev was crushed between reformers and reactionaries but, after defeating hardliners in Moscow, Yeltsin spread a map of reform across Russia which enabled free trade systems to grow. These volatile times also encouraged cronyism, corruption and organised crime fed by money laundering which international electronic banking could accomplish with speed and secrecy. The financial crash in Russia in 1998 damaged the prospect of a fully westernised state of Russia in the short term while unrest and hostilities were at their most devastating in Russia's wars with Chechnya which have continued into the new century.

Just as Russia's turbulence was apparent at the start and finish of the 20th century so was Serbia's. Yugoslavia dissolved as Bosnia, Croatia and Slovenia declared independence. Ethnic conflict in Bosnia, where 200,000 died, was the first major war in Europe since 1945. The shattered city of Sarajevo became an image of the religious and racial divisions which were rekindled in the Balkan states with war also in Kosovo and traditional enmity between Croat and Serb re-emphasised. The changing face of post-1945 Europe was confirmed by the reunification of Germany in 1990 and the Czech and Slovak republics forming separate states in 1993. Even the power of the German economy, however, found difficulty with the extent of economic rebuilding needed in its eastern lands.

At the end of the century the world's wars were smaller but just as vicious and civil wars brought genocide and ethnic cleansing, with particularly horrific loss of life in Rwanda where 2 million people were estimated to have also left the country. The most powerful nations were relatively successful in containing strife in some parts of the world but much less so in others. United Nations or NATO involvement in Bosnia, Kosovo, Somalia—though preventing worse crises—brought controversy or was less effectively accomplished than distressed peoples needed. Apart from Rwanda's agony savage fighting in such countries as Afghanistan (with Islamic fundamentalism resurgent), Liberia and East Timor, atrocities against the Kurds, famine in Ethiopia and floods in Mozambique at the century's turn, could not be stopped or were not helped with the speed the world wanted. But there were increasing signs that nations would band together to oppose aggression against neighbours by rogue or unstable regimes. Iraq's invasion of Kuwait in 1990 was overturned in 'Operation Desert Storm' by a coalition of the USA, European and Arab nations. Allied bombing of Iraq and of Serbia during the Kosovo crisis relied on overwhelming use of missiles and 'stealth' attack aircraft as military planners aspired to wars of modern technology, against pinpointed targets, which would not lead to huge troop casualties.

President de Klerk saw that South Africa could not survive with the social and economic policies under which it had lived for so long. He boldly sanctioned the undoing of the legal and racial apparatus of apartheid with the support of the newly-freed Nelson Mandela. A peaceful transition from white to (largely) black rule took place with the founding of the 'rainbow republic' and the election of Mandela as its president. Mandela's own courage and his magnanimity in victory gave him a moral standing little seen since the days of Mahatma Gandhi. Though it was recognised that the USA could no longer be expected by many to be the world's policeman its influence helped particularly two moves towards peace in the Middle East and Europe. Hopes of reasonably peaceful coexistence began when Palestinian authorities took over responsibilities from Israel in such areas as Jericho and Gaza and the long-opposing sides in Northern Ireland reached an historic, if fragile, Good Friday Agreement.

The booming US economy, with computer giants such as Microsoft and Oracle, fuelled one of the most sustained periods of the USA's economic growth. Parts of Europe shared this prosperity. Not even the threatened impeachment of Bill Clinton and scandal in the White House dented the USA's around 6% growth each year in the later 1990s. This optimism was reinforced by fresh trade agreements and by new political or financial structures like the European Bank of Reconstruction and Development to regenerate Eastern Europe. The Maastricht Treaty coming into force in 1993 brought greater European political and economic co-operation, forming the European Union and a single market. A single European currency, the euro, was later introduced. But while leading economies surged ahead it was estimated that 40% of the global economy was in recession by 1999. Japan's post-war economic growth came to a standstill helping to bring economic hazard to other countries in South-east Asia including an economic slowdown in Hong Kong which had transferred into Chinese hands after over 150 years of British rule. Prices of oil and raw materials fell across the world.

International banking became ever more coordinated but so did sophisticated scope for clever financial manipulation as testified by the Russian mafia's criminal activities and the spectacular crash of the BCCI bank through fraud. There were riots in Seattle during the World Trade Organisation's meeting. These protests were against unchecked free trade and those policies of the World Bank and the International Monetary Fund which many young demonstrators perceived as encouraging tax increases and spending cuts even in the poorest countries. They favoured writing off Third World debt and closing the gulf between rich and poor. The Vietnam generation had protested about war. Now their children protested about trade and for much greater coordinated aid to be given to the Third World in line with what many viewed to be Keynes' economic vision which had helped revive Western economies decades earlier.

By the 1990s applications in space, electronic and transport technologies had been transformed beyond recognition compared with the 1950s. The Hubble telescope ferried into orbit by space shuttle captured pictures of galaxies in their infancy, the COBE satellite helped confirm the Big Bang theory of the universe's origin and astronomers located the most distant object in space (a quasar) yet seen. General Motors made a breakthrough with the first mass-market electric car and the *Thrust SSC* car broke the sound barrier. But if there were keystones to the inventions of the century's end they might be found in convergence and transfer among technologies and within genetic investigation, in man's interaction with the machines he built as well as his growing awareness of links with the animal, bird and insect worlds, in his creation of life and understanding of the structure of mind and body.

'Dolly' the sheep was cloned by nuclear cell transfer, a computer defeated the world champion (Kasparov) in a chess game, the genes responsible for mental handicap and dyslexia were discovered and genetic fingerprinting of living relatives proved that long-buried bones were those of the last Russian Tsar and his family. Videophones were marketed and voice controlled computers were used for the first time to accomplish hands-free surgery to improve the speed and safety of operations. William Dobelle announced bionic eye glasses to help restore sight and drugs were developed to battle degenerative brain disease. Following research on rats, limbs and sensation in them were revived after human spinal injuries while analysis of cancer's causes discovered—with many implications for future research—that fruit flies share 60% of human genes and biochemical pathways.

A first summit meeting on the environment was convened. But there was an alarming omen of future dangers when global warming, due to atmospheric disruption from human activities and energy consumption, caused the Prince Gustav Ice Shelf and Larsen Ice Shelf in Antarctica to begin to disintegrate. Construction technology, however, flourished in the 1990s when the Channel Tunnel between Britain and France and the world's tallest building in Malaysia were both opened and the new Guggenheim Museum in Spain was designed, using technology that had been developed first for the aerospace industry.

As the year 2000 approached 60 million people world-wide were joining the

Internet each third of a year but hackers were engaging too in electronic terror-ism via computer viruses. There was speculation that business through electronic technology ('e-commerce') would eliminate the roles of many middlemen and that the patents system, which had served innovation so well in the past, might be used now to stifle 'free' Internet progress. It was estimated that 85% of the Internet's data pages were currently in the English language. Thus it would be appropriate to record the scale of man's inventions between 1900 and 1999 not in the Latin of the epitaph to the great architect, Christopher Wren, but in its English version:

If you seek his monument, look around.

The Adidas Predator™ boot

Football boot giving greater accuracy and control in receiving and passing the ball
Craig Johnston, Sydney, Australia for Zermatt Holdings, London, England
Filed 19 June 1991 and published as WO 92/22224, EP 544841B and US 5437112

This invention is an example of consumerism by the young. Craig Johnston was born in Johannesburg, South Africa in 1960. His family moved to Australia and he then went to England to play association football, or soccer, for Middlesbrough and Liverpool. He retired from football at the age of 27 and went back to Australia to care for his sister, who had been poisoned by gas in an accident.

Johnston is said to have thought of the boot when he noticed a rubber bung falling. The illustrated boot comes from an example of a well-written patent specification. Johnston knew that it was often difficult to control the ball properly, particularly in wet conditions, where the player might slip. Boots for kicking have a 'sweet spot' for the best contact, just like tennis rackets. The patent specification claims that on conventional boots it is not easy to identify the correct area for striking the ball. The general idea is to increase the surface area in contact with the ball to give better control. Balls are usually received on the upper part of a boot, which is normally convex, reducing the surface area.

On this boot, the ribs shown on the figure at 70 deform under the impact of a ball to increase the surface area affected, and hence control the direction, and increase the speed, of the ball. The ribs are made of an elastomeric material to increase friction. A gloving effect is also produced which keeps the received ball more stable and makes it easier to control. The ball is normally struck when passing or shooting by using the side of the foot and the figure shows additional ribs 72, which are grooved to help them deform easily, again increasing contact with the ball. The patent application explains that 55 at the bottom of the figure is covered by an earlier application, WO 91/11929.

WO 92/22224 contains a search report listing previous documents covering the same subject area. The six given date back to 1960 and include two extremely short British documents which were never given protection, but which do mention the idea of improving contact with the ball. Many modern patent documents contain search reports which are useful sources of related material. Search reports often reduce the coverage of any granted patents, as in this case, where the number of claims (the monopoly requested) was reduced from 27 in WO 92/22224 to 14 in EP 544841. The boot was launched as a product as the Adidas Predator™ boot. Johnston in fact later joined a rival firm, Reebok, as head of product and marketing. Six thousand teams play in the English Schools' Football Association Adidas Predator Cup.

FIG.7(b)

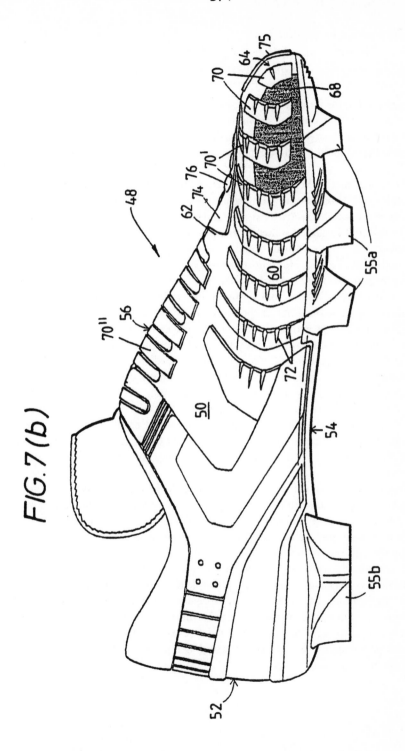

'Clockwork' radio

Radio powered by winding it up
Trevor Graham Baylis, Twickenham, Middlesex, England
Filed 19 November 1991 and published as **GB 2262324**

Trevor Baylis was born in 1937. He had a background in engineering and in physical education, especially in swimming, in which he competed for a time. He was also a stuntman. Later he ran a business developing and building swimming pools. He also experimented with ideas in his workshop.

One night he was watching a television programme about AIDS in Africa. The message was that if the people were better informed there would be a less drastic mortality toll. Broadcasting was of relatively little use as, even if people could afford radios, there was no electricity supply and batteries were expensive. Baylis thought about this and imagined someone in a pith helmet listening to a wind-up gramophone player. Surely a spring could be wound up so that the power compressed in it could power a radio. He went to his workshop and found a direct current motor which he ran in reverse to create a dynamo. This was attached to the battery terminals of an old transistor radio. He wound the dynamo with a hand brace and created a sound. He knew then that the principle would work.

He experimented with a number of designs, including a pulley hanging from a tree. Then he came up with a combination of a spring and a cog wheel producing a radio that could work for 14 minutes before it needed 2 minutes' of rewinding. By cranking a handle, metal wire is wound tightly onto a drum. As it unwinds in use voltage is created via a gearbox leading to a generator. He then applied for a patent and began to talk to manufacturers, who dismissed the idea. The BBC television programme *Tomorrow's world* filmed him demonstrating the radio and this did result in a response. South African-born accountant Christopher Staines expressed an interest, and the South African Liberty Life Group got a company going with help from others.

Baygen began production of the Freeplay® radio in 1994 at a factory in Capetown. Disabled people are recruited and are paid the same wages as able-bodied workers. 120,000 units are made monthly and are shipped to such places as Kosovo as well as in Africa. An important market is hikers in North America who want to keep in touch with the outside world. The cost is still relatively high at about £69 in Britain. In 1997 a smaller and cheaper version was made for the Western market.

Baylis has suggested carrying the principle on to torches, lap top computers and other applications. Meanwhile he is devoting much energy to a Royal Academy of Inventors, a concept involving helping inventors getting their ideas to the market.

The principle is in fact not new, and five patents were cited against it in the British patent application, which was withdrawn in 1996. The earliest of these is GB 217492, filed in 1923.

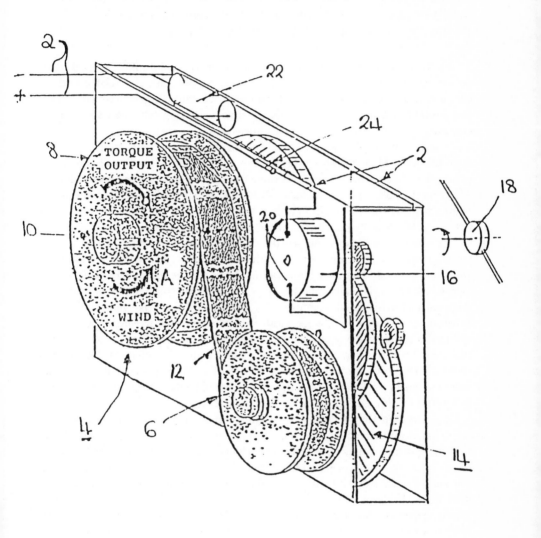

TORQUE
OUTPUT

WIND

Cloning animals

Cloning animals by implanting altered denucleated cells into surrogate mothers
Keith Campbell and Ian Wilmut for Roslin Institute, Edinburgh, Scotland
Filed 31 August 1995 and published as WO 97/07668-9

This is the famous 'Dolly the sheep', a highly controversial invention. Ethical issues do not normally come into assessing inventions, but here concerns related to those over genetically modified food have been raised. Wilmut is a geneticist and Campbell a cell cycle biologist. The experiment involved taking cells from adult sheep, culturing them *in vitro*, transferring the nucleus to an ennucleated egg implanted in a surrogate mother. This resulted in a lamb identical to its genetic mother, just like identical twins. Dolly the sheep was born on 5 July 1996, the only 1 of 29 implantations to succeed, which themselves came from 277 eggs which had been cultured. The patent applications were filed before the actual birth, and the official announcement was only made in February 1997, just before the publication of an article in *Nature* and the publication of the patent applications.

The *Observer* newspaper jumped the gun in releasing the story. There was intense interest, not all complimentary, although the singer Dolly Parton was apparently very pleased. There are a number of claims for the technology being a good idea. Cloned animal organs, especially pigs, could be transplanted into humans, and the quality of herds could be improved. Clones would also make good subjects for experiments as they would be identical. Parents who feared passing on a defect to children could have the fertilised ovum cloned and the duplicate tested, so that if the clone were free from defects then the other could be implanted. Duplication could also be carried out to help with *in-vitro* fertilisation. Supporters also claim that even if a person were cloned then the second would be like a delayed identical twin and would be different because of environmental factors.

Opponents cite the concept as being unnatural and unethical; playing God; having to pay royalties on offspring; and a fear about where the technology would lead with evil people cloning themselves and cloned humans being farmed for 'spare parts'. It has also been suggested that eventually a new race could evolve which was free from defects and with superior qualities who would not be able to interbreed with human beings.

The Geron Corporation, an American company which only started up in 1990, merged with Roslin Bio-Med, part of the Roslin Institute, to form Geron Bio-Med, the headquarters of which is at the Roslin Institute. The patent rights belong to this new company. The first commercial use of the technology was in the second lamb to be born through the technique, Polly, which had a human gene so that her milk included human Factor IX, a blood-clotting agent needed by men with haemophilia B. Dolly herself had a normal lamb in due course, Bonnie. Sheep were chosen partly because they as mammals have a certain resemblance to humans and, perhaps facetiously, because they are 'very cheap' in Scotland.

1

UNACTIVATED OOCYTES AS CYTOPLAST RECIPIENTS
FOR NUCLEAR TRANSFER

This invention relates to the generation of animals
including but not being limited to genetically selected
and/or modified animals, and to cells useful in their
generation.

The reconstruction of mammalian embryos by the transfer
of a donor nucleus to an enucleated oocyte or one cell
zygote allows the production of genetically identical
individuals. This has clear advantages for both research
(i.e. as biological controls) and also in commercial
applications (i.e. multiplication of genetically valuable
livestock, uniformity of meat products, animal
management).

Embryo reconstruction by nuclear transfer was first
proposed (Spemann, *Embryonic Development and Induction*
210-211 Hofner Publishing Co., New York (1938)) in order
to answer the question of nuclear equivalence or 'do
nuclei change during development?'. By transferring
nuclei from increasingly advanced embryonic stages these
experiments were designed to determine at which point
nuclei became restricted in their developmental
potential. Due to technical limitations and the
unfortunate death of Spemann these studies were not
completed until 1952, when it was demonstrated in the
frog that certain nuclei could direct development to a
sexually mature adult (Briggs and King, *Proc. Natl. Acad.
Sci. USA* **38** 455-461 (1952)). Their findings led to the
current concept that equivalent totipotent nuclei from a
single individual could, when transferred to an
enucleated egg, give rise to "genetically identical"

Convection towers to cleanse the air

Tall towers which use water sprays to cleanse the air of pollution
Melvin Prueitt, Los Alamos, New Mexico for the Department of Energy
Filed 9 September 1992 and published as WO 94/05398 *and* US 5284628

There have been many attempts at, and suggestions for, cleansing the air, and this must serve merely as an example, as not one has (yet) come into use. Melvin Prueitt, a 'guest' physicist at the Los Alamos National Laboratory, has patented one such invention. He points out that filters would be expensive and would require a huge throughput of air, and proposes using natural forces. The patent was the result of a Department of Energy contract.

The tall and hollow tower, made of a steel framework with teflon-coated fibre-glass stretched round it, has water sprays at 13 and 14. These cool and humidify the air coming in from the top, causing it to sink and hence draw in more air. At the bottom any water left and particles from the air are collected in catch basins. The air runs through turbines to generate power and, after a final water spray, the air leaves through numerous tunnels at ground level so that it displaces polluted air upwards. The result is air that has been cleansed, cooled and humidified, which perhaps is not ideal for climates such as England's. The water can be seawater, or cleansed sewer water.

Prueitt says in the patent that computer calculations show that in a climate of 30 °C and a humidity of 50%, a tower 500 metres high and 200 metres wide would process 121 cubic kilometres of air per day. The air would flow downwards at 40 kilometres per hour. The turbines would generate 85 megawatts and the pumping requirement for the sprays would be 60 megawatts. Less power would be generated in less humid conditions, and the towers need to be tall to generate a surplus. 15,000 kilogrammes of water per second would be required. Elsewhere, Prueitt says that 200 such towers, 200 metres high, would be enough to cleanse Los Angeles at a construction cost of about $2 billion.

Prueitt is the only scientist in a family of artists, and he combined these two influences when, having to write software programmes in the 1970s at the Los Alamos Laboratory, he tried making three-dimensional works of art out of huge sets of numbers. He is now popular as an artist making vivid colour images created by computer programmes, many of which have been published in *National Geographic* and other magazines. Prueitt has two companies, Solar Energy Ltd and Hydro-Air Technologies. He continues to work on his ideas. US 5441238 is a heat engine using brine. WO 99/46545, an 'Air conditioner energy system', is a modification of his ideas for a smaller, horizontal device which can be used in warm, dry climates to provide cool air and electric power, which he calls ACES. The basis of the idea itself may lie in Phillip Carlson's US 3894393, filed in 1974, which envisages tunnels along steep mountain slopes.

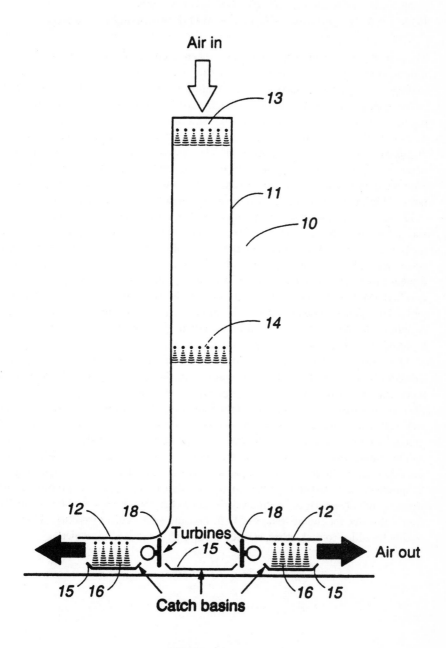

Fig. 1

Foldable portable buildings

Folding buildings for ease of moving and construction
Juan Kalinowski, Langley, British Columbia, Canada
Filed 3 February 1995 and published as WO 98/02626 *and* US 5596844

Mass-production techniques have rarely been used in building construction, although there are some examples from the 1940s in the USA. It is expensive, inefficient and slow to construct each building from many materials, and doing so in the open air can cause problems because of bad weather. It is also awkward moving the materials to the construction site. An example of a solution to this problem is illustrated. The inventor was born in Peru and trained as an engineer, specialising in construction and transport work. He was told to build housing for maintenance workers on the Peruvian railways at an altitude of 5,000 metres. He had always enjoyed sketching houses as a child, and now he could see the practical problems of building in a hostile environment. Cement hardly sets at such heights.

He spent ten years perfecting his ideas, helped by his wife, an architect. He wanted to copy the concept of computer files which are compressed while being moved electrically. 'We cannot take the luxury of shipping air from the plant to the other end'. The construction is of steel bonded to rigid polystyrene insulation. The dimensions when extended are about 5.4 metres long, 2.4 metres wide and 2.6 metres high. Only the long parallel walls are load-bearing, and they include windows and a door. When folded up, it shrinks into a sixth of the compact size. The weight is about 6.6 tonnes. A team of three unskilled workers are meant to construct a building in six hours with the aid of in-built hand-winches.

The central portions of the building (floor, walls and roof) are rigid and everything folds out from there. The end floors form the remaining sides. Packed up, it meets international standards for transporting as its own container. This means that there is no wasteful packaging. The end floors drop down from the centre, then the end walls and roofs extend from the centre floor and roof as two wall sections also gradually unfold. The illustration shows this process almost finished.

The roof consists of a ceiling with trusses between it and the roof (which rests on the side walls). The trusses are hinged so that they are swung up to support a gently pitched roof. Partition walls can be installed if required, but plumbing and wiring is already provided. Kalinowski believes these buildings would be invaluable after a disaster for the provision of both housing and medical facilities. They can readily be added to each other, stacked three high, or moved by trailer.

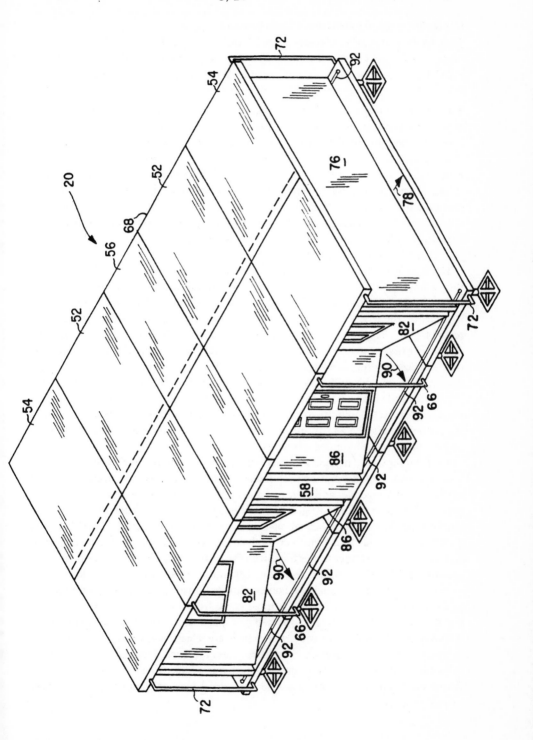

Fuel cells

Obtaining power by combining hydrogen and oxygen
Jingzhu Wei, Burnaby, Charles Stone, Vancouver and Alfred Steck, West Van-
couver for Ballard Power Systems, Burnaby, all in British Columbia, Canada
Filed 21 September 1993 and published as WO 95/08581 *and* US 5422411

This invention really dates back to 1949 and Francis Bacon's GB 667298, but it is only in the last few years that costs have come down enough for this to be a genuinely exciting potential source of power. Instead of using internal combustion engines, chemistry is used to provide power for either vehicles or stationary motors which could, for example, power a home. A combination of hydrogen and oxygen results in emissions of water vapour as the only pollutant. Internal combustion engines are only 15 to 18% efficient, while fuel cells are 40 to 60% efficient. Fuel cells are also silent. These factors would make fuel cell-powered vehicles a great boon in crowded neighbourhoods. If the idea takes off, then battery-powered electric cars may be unable to compete in the market.

Because of high costs, fuel cells have been relatively little used so far, although they were used by NASA in the Gemini spacecraft in the 1960s as a source of both electricity and drinking water through its water by-product. Ballard Power Systems is probably the leader in bringing costs down so that it can become a widespread technology. They have numerous patents covering different aspects of an ion-exchange membrane for use in an electrochemical fuel cell, and this patent application will serve to represent the technology. What happens is that an electrode has hydrogen on one side and oxygen (from the air) on the other. Catalysts on each side of the electrode ionise both the hydrogen and the oxygen. A membrane allows the ionised hydrogen to join the oxygen on the other side of the electrode to form water (H_2O), with energy being a by-product of the reaction. Numerous fuel cells are used to form appreciable amounts of energy.

Ford Motor Company and DaimlerChrysler have both invested hundreds of millions of dollars in collaboration with Ballard. Successful trials have been carried out with buses in Vancouver and Chicago, and it is hoped that the first cars will be manufactured by 2004. The hydrogen itself can be extracted from different kinds of fuel within a factory, and then carried on-board as hydrogen, or it can be provided from a feed-stock as a gas by a 'reformer' on board a vehicle. Without reformers hydrogen must be stored either under high pressure or at low temperatures, both of which are awkward for automobiles, and may explain the trials with the larger buses.

Although fuel cells as such are non-polluting, Canada's Pembina Institute points to a major problem. The hydrogen must come from somewhere in the first place, and the method of extracting it can itself be a major source of greenhouse emissions. The leading contenders for the original fuels are 'clean' petrol and methanol, both of which would lead to greenhouse effects. Less favoured, but much less polluting, is taking the hydrogen from natural gas, which is a bulky fuel to carry on a vehicle. Better still would be to strip hydrogen from fossil fuels or even water by using renewable energy.

TRIFLUOROSTYRENE AND SUBSTITUTED TRIFLUOROSTYRENE COPOLYMERIC COMPOSITIONS AND ION-EXCHANGE MEMBRANES FORMED THEREFROM

Field Of The Invention

The present invention relates to trifluorostyrene based polymeric compositions.
More particularly, the present invention relates to
5   polymeric compositions derived from copolymers of α,ß,ß-trifluorostyrene with a variety of substituted α,ß,ß-trifluorostyrenes. These compositions are particularly suitable for use as solid polymer electrolytes in electrochemical
10   applications, such as, for example, electrochemical fuel cells.

Background Of The Invention

A variety of membranes have been developed over the years for application as solid polymer
15   electrolytes for fuel cells and other electrochemical applications. These polymers have typically been perfluorinated aliphatic compositions, such as those described in U.S. Patent Nos. 3,282,875 and 4,330,654. These
20   compositions are very expensive membranes, and in the case of the '875 patent tend to exhibit poor fuel cell performance characteristic at high current densities. Alternatively, a series of low-cost polyaromatic-based systems have been in-
25   vestigated, such as those described in U.S. Patent Nos. 3,528,858 and 3,226,361. These materials suffer from poor chemical resistance and mechanical properties which tend to limit their use in fuel

Fullerenes

New structures of carbon molecules
Donald Huffman, Tucson, Arizona and Wolfgang Kratschmer, Gaiberg,
Germany for Research Corporation Technologies, Tucson, Arizona
Filed 30 August 1990 and published as **WO 92/04279**

Fullerenes evolved as a result of a 1985 experiment on carbon molecules. The patent specification is for the first method to make quantities of the substance (and was quickly superseded by further research) and not for the discovery of the substance.

Carbon atoms are good at attaching themselves to other atoms, and can be used to form long chains called polymers, which are used in many products such as plastic cups and bottles. An experiment was carried out at Rice University at Houston, Texas by Harold Kroto of Southampton University, England and Richard Smalley and Robert Curl of Rice. They wanted to simulate conditions near the surface of a star to see how large molecules in space might be formed. They fired an intense pulse of laser light at a carbon surface in the presence of helium (hydrogen and nitrogen, then just nitrogen, were initially tried). Gaseous carbon then combined with the helium to form clusters. After cooling the gas to near absolute zero, the most abundant clusters in the spectral analysis turned out to be C60.

C60 means 60 atoms of carbon in a single molecule. The scientists had never seen anything like it. There was a structured, spherical form which reminded them of Buckminster Fuller's geodesic dome, which they had seen in the shape of the American pavilion at the 1967 Montreal Expo (see page 130). They were unable to recreate it on a computer so had to resort to scissors, paper and tape to do so. They decided to call the new structure buckminsterfullerene, but as that is rather a mouthful, it was shortened to its accepted name, fullerenes. Soccerene was briefly considered as an alternative name. They are also often nicknamed buckyballs. It was interesting—but could any practical use be made of the structure?

A strange property in fullerenes is that some of the electrons in the carbon atoms are 'de-localised' so that they don't seem to realise that they are part of the structure. This means that other atoms can easily be added to make, for example, superconductors or insulators. In 1991 the topic exploded as masses of papers were written about fullerenes. In 1996 the three scientists were jointly awarded the Nobel Prize for Physics.

Although the field is still quite new, one promising idea seems to be that of altering the structure to form 'buckytubes'. These are thin hollow fibres which have 200 times the tensile strength of steel. These can form tiny tweezers for picking up clusters of molecules, or containers for delivering minute quantities of drugs, or shields against radioactivity. Alternatively they could be made into cages which hold some molecules while letting others of a smaller size pass through. By adding other kinds of atoms particular qualities can be provided such as measuring electrical resistance.

FIG.I

Programmable materials

Robot structures controlled by software which alter shape
Joseph Michael, London, England
Filed 4 March 1994 and published as WO 95/23676 *and* GB 2287045B

'Programmable materials' is the title of Joe Michael's invention. It is also called a 'shape-changing robot' and 'fractal robot'. Born in 1962, he is a software engineer. By using a mixture of robotics and software, structures are enabled to change shape if that is desirable. Traditionally machines have specific functions in specific positions, but that limits the potential utility. For example, a bomb-defusing device could change shape to crawl through a tunnel and then form a wall round the bomb before defusing it. The idea is that there are identical cubes or 'monomers' which pass on software instructions or form fastenings by using ports in the middle of each face. Most of these cubes are hollow. Each monomer has a micro-controller, motor, relays and sensors. Figure 1 shows possible arrangements, while the right-hand Figure 2 shows one that is able to walk. The bottom figures show monomers pushing each other along on (unshown) 'streamers'. Movement is obtained by their sliding in relation to each other.

Although each monomer has identical features they could also be given special features for special uses, such as the ability to handle delicate or small objects, or to carry tools or other equipment. Instructions could be given for particular purposes such as entering a nuclear reactor and carrying out activities in a radioactive environment. If a monomer were faulty or damaged a new one could be supplied from a reserve store, and inserted in place by a functioning monomer. It has been suggested that very small models could carry out operations within patients. In fact the possibilities for their use and for special features are almost endless.

It is suggested that the factory for making such cubes could itself be made from such cubes. Factories (or rather, the assembly lines) making a variety of consumer products could be made from such cubes, with quick and easy adaptations when production of one product or model changes to another. This would be environmentally friendly as there would be no waste as the cubes could simply be reprogrammed. Michael also suggests their use in a hostile environment such as space. Michael has set up his own company, Robodyne Cybernetics, and has written many papers on aspects of developing his invention. Now he needs a great deal of finance. It is true that WO 95/23676 quotes in its search report EP 129853, a Hitachi filing dating from 1983, for 'cellular robots', which has been granted protection. This has considerable similarities and may perhaps deserve to be considered the first of its kind.

1/19

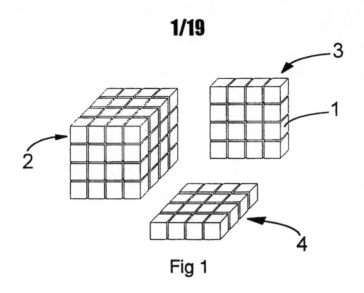

Fig 1

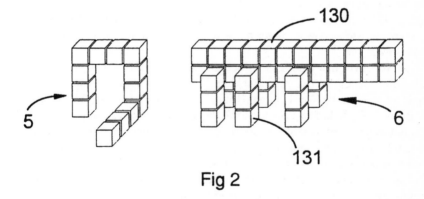

Fig 2

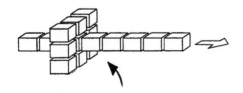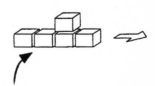

Pyramidal teabags

Method of making teabags in the shape of a pyramid
Kevin Fincham, David Seaward, Graham Shirley and Geoffrey Vernon for
Molins plc, Coventry, Warwickshire, England
Filed 8 July 1993 and published as WO 95/01907 *and* US 5548947

No book published by a British author on inventions would be complete without a mention of tea. The idea of bags to hold tea leaves for infusion has been around since at least 1918, when Benjamin Hirschhorn filed a patent application (in the USA, in fact, with US 1310796). This was a small sack sealed at one end, and the classic flattish, square bag emerged later. Unilever plc worked on the idea of a new kind of teabag for four years. They hoped that it would improve the flavour but also wanted to attract a new and younger audience to tea drinking, which had an old-fashioned image. Their Brooke Bond PG Tips brand was well known and popular but sales were static.

The drawing shows the way the pyramidal bags are made. The web of packaging unwinds off a roller and forms a tubular shape round the tube. The tea leaves are delivered down the tube itself, with the bags sealed and separated from the web as 'each dose of filling material [tea leaves] is delivered'. A valve controls the dosing at the exit of the tube, marked 24a. The sealing and severing is done by the rotors on all four sides of the tube, the arms of which 'interdigitate to make successive seals in mutually transverse directions, so forming tetrahedral packets'. Such an invention was of course essential in order for a low-cost product like teabags to be made easily and cheaply in large quantities. The work of making a cheap fabric which held the leaves in while allowing water to move through had of course been done many years before.

Unilever spent many millions of pounds on a publicity campaign for their new teabags, and sales did rise. The publicity included 20 metre high giant pyramids being erected simultaneously in ten British cities, with the slogan 'Much ado about pyramids' being linked to a Shakespeare theme, and activities being provided. The pyramid structure allows 50% more room for leaves to move round than the conventional teabags. The company claims that this ensures a fuller taste, similar to that from a teapot. Others feel that all it does is to speed up the process of infusion. Unilever moved to protect other aspects besides the patent. Two British registered designs, 2035706 and 2043840, protect the appearance of the bag, while the word Pyramid® and the phrase More room to move® have been registered as trade marks.

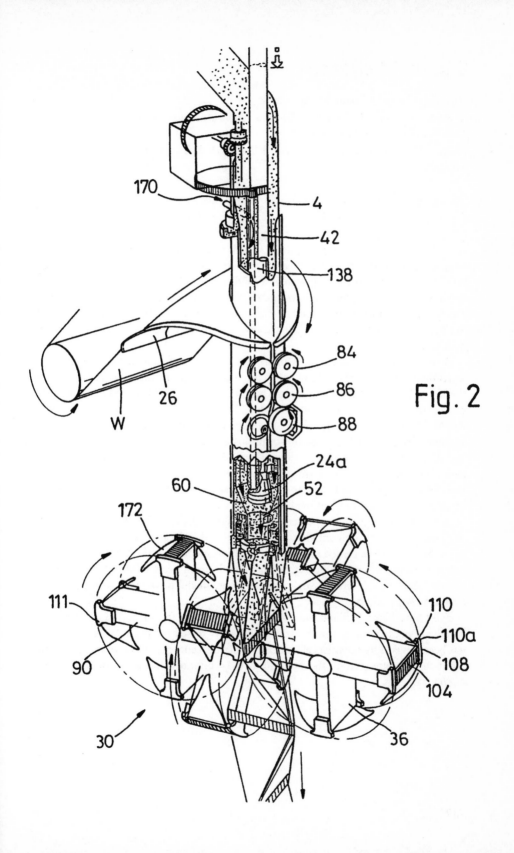

Fig. 2

Sildenafil citrate

Pharmaceutical dilating blood vessels
Andrew Simon Bell, David Brown and Nicholas Kenneth Terrett for Pfizer
Ltd, Sandwich, Kent, England
Filed 20 June 1990 and published as EP 463756, US 5250534, 5346901 *and*
5791283

This pharmaceutical has caused much hilarity, although it is helpful for sufferers from a common malady. It is a pyrazolopyrimidinone antianginal agent which means that it helps to cope with angina, hypertension, heart failure and atherosclerosis. It does this by dilating the blood vessels so that more blood can flow. Judging from the patent literature it took some time for the penny to drop that this might be a useful treatment for impotence. The filing of WO 94/28902 on 9 June 1993 disclosed that possible use as it was described as a 'medicament for the curative or prophylactic treatment of erectile dysfunction in a male animal, including man'. EP 812845, by two other Pfizer scientists, discloses an effective way to make the pharmaceutical.

Better known as Viagra®, this pharmaceutical was at first illicitly imported in small quantities by air travellers before, in March 1998, it was approved for use in the USA. Years of clinical studies had been carried out but Pfizer did not realise the impact that would be made on lifestyles. The chairman, Bill Steed, said that he would fall asleep before the television to one comedian's jokes only to wake up to another's. Both the number of prescriptions and the share price of Pfizer shot up, and for a time it was the fastest-selling (legal) drug in history. The advertising campaign included Bob Dole, former Republican presidential candidate. There were worries in Britain, though, that the liberal prescribing of this pharmaceutical would break the budget for the National Health Service.

Some deaths were also reported among users. This is mainly because its raising of blood pressure is a potential hazard for those who already have high blood pressure, although there can be other risks. By November 1998 the US Federal Drugs Administration was aware of 155 deaths, mostly from heart failure. Users can also have unpleasant side-effects such as headaches, flushing, stomach aches and mild problems with vision.

Sales have since declined but consumer recognition for Viagra® is phenomenal. There are many sites on the Internet which offer to sell the product. A nasal spray has been devised by scientists at the University of Kentucky which reduces the time for it to take effect from one hour to 5 to 15 minutes. It was published as WO 99/66933.

This invention relates to a series of pyrazolo[4,3-d]pyrimidin-7-ones, which are potent and selective inhibitors of cyclic guanosine 3',5'-monophosphate phosphodiesterase (cGMP PDE), having utility in a variety of therapeutic areas including the treatment of various cardiovascular disorders such as angina, hypertension, heart failure and atherosclerosis.

5 The compounds of the invention exhibit selectivity for inhibition of cGMP PDEs rather than cyclic adenosine 3',5'-monophosphate phosphodiesterases (cAMP PDEs) and, as a consequence of this selective PDE inhibition, cGMP levels are elevated, which in turn can give rise to beneficial platelet anti-aggregatory, anti-vasospastic and vasodilatory activity, as well as potentiation of the effects of endothelium-derived relaxing factor (EDRF) and nitrovasodilators. Thus the compounds have utility in the treatment of a number

10 of disorders, including stable, unstable and variant (Prinzmetal) angina, hypertension, congestive heart failure, atherosclerosis, conditions of reduced blood vessel patency e.g. post-percutaneous transluminal coronary angioplasty (post-PTCA), peripheral vascular disease, stroke, bronchitis, chronic asthma, allergic asthma, allergic rhinitis, glaucoma, and diseases characterised by disorders of gut motility, e.g. irritable bowel syndrome (IBS).

15 European patent application EP-A-0201188 discloses certain pyrazolo[4,3-d]pyrimidin-7-ones as adenosine receptor antagonists and PDE inhibitors, useful in the treatment of cardiovascular disorders such as heart failure or cardiac insufficiency. However these compounds are neither particularly potent PDE inhibitors, nor are they claimed to be selective inhibitors of cGMP PDE.

 The compounds of the present invention are of the formula (I):

20

(I)

wherein

40 R^1 is H, C_1-C_3 alkyl, C_3-C_5 cycloalkyl or C_1-C_3 perfluoroalkyl;

 R^2 is H, C_1-C_6 alkyl optionally substituted by OH, C_1-C_3 alkoxy or C_3-C_6 cycloalkyl, or C_1-C_3 perfluoroalkyl;

 R^3 is C_1-C_6 alkyl, C_3-C_6 alkenyl, C_3-C_6 alkynyl, C_3-C_7 cycloalkyl, C_1-C_6 perfluoroalkyl or (C_3-C_6 cycloalkyl)-C_1-C_6 alkyl;

 R^4 taken together with the nitrogen atom to which it is attached completes a pyrrolidinyl, piperidino, morpholino, or 4-N-(R^6)-piperazinyl group;

45 R^5 is H, C_1-C_4 alkyl, C_1-C_3 alkoxy, NR^7R^8, or $CONR^7R^8$;

 R^6 is H, C_1-C_6 alkyl, (C_1-C_3 alkoxy) C_2-C_6 alkyl, hydroxy C_2-C_6 alkyl, (R^7R^8N)C_2-C_6 alkyl, (R^7R^8NCO)C_1-C_6 alkyl, $CONR^7R^8$, $CSNR^7R^8$ or $C(NH)NR^7R^8$;

 R^7 and R^8 are each independently H, C_1-C_4 alkyl, (C_1-C_3 alkoxy)C_2-C_4 alkyl or hydroxy C_2-C_4 alkyl;

 and pharmaceutically acceptable salts thereof.

50 In the above definition, unless otherwise indicated, alkyl or perfluoroalkyl groups having three or more carbon atoms may be straight or branched chain. In addition alkenyl or alkynyl groups having four or more carbon atoms, or alkoxy groups having three carbon atoms, may be straight or branched chain.

 The compounds of formula (I) may contain one or more asymmetric centres and thus they can exist as enantiomers or diastereoisomers. The invention includes both mixtures and separate individual isomers.

55 The compounds of formula (I) may also exist in tautomeric forms and the invention includes both mixtures and separate individual tautomers.

 Also included in the invention are radiolabelled derivatives of compounds of formula (I) which are suitable for biological studies.

FURTHER READING

THIS book has many, many sources. The patent specifications held at the British Library provided much information, of course, but the others cannot be listed as they are too many. One thing that was noticed was how often different sources contradicted each other, and hard choices often had to be made in what are after all short and simple accounts. It is worthwhile looking at different sources in any research for this reason if possible, as the facts and detail often vary considerably. Primary sources such as archives are better then second-hand accounts, of course, but are usually impossible to identify.

A number of books are useful for research into the history of inventions. Some are listed below, followed by a mention of sites on the Internet and of the patent libraries that can be visited in Britain and the United States. The only ones listed which regularly mention patent numbers are *New and improved* and *Historical first patents*. They both include extensive reproductions from the patents themselves.

Books arranged by broad topics

The book of inventions and discoveries 1992. V-A Giscard d'Estaing. 1991, Macdonald.
The evolution of useful things. P. Petrovski. 1993, Knopf.
The new Shell book of firsts. P. Robertson. 1994, Headline (includes an index by place).

Books arranged alphabetically by invention

The Guinness Book of innovations: the 20th century from aerosol to zip. G. Tibballs. 1994, Guinness.
Historical first patents: the first United States patent for many everyday things. T. Brown. 1994, Scarecrow Press.
New and improved: inventors and inventions that have changed the world. R. Baker. 1976, The British Library (gives patent numbers, British mostly, for 363 inventions).
They all laughed . . . from light bulbs to lasers: the fascinating stories behind the great inventions that have changed our lives. I. Flatow. 1993, Harper Perennial.

Books arranged chronologically by invention

The Harwin book of inventions, innovations, discoveries from pre-history to the present day. K. Desmond. 1987, Constable (no detail, but a very good index by specific topic so a good way to identify the inventor and the date).

Other categories

British patents of invention, 1617–1977: a guide for researchers. S van Dulken. 1999, The British Library.
How to find information: patents on the Internet. D Newton. 2000, The British Library.

Introduction to patents information. Edited by S van Dulken. 3rd ed., 1998, The British Library.

Inventors at work. K. A. Brown, 1988, Tempus Books (interviews with prominent American inventors).

Rooms near Chancery Lane: The Patent Office under the Commissioners, 1852–1883. J. Hewish. 2000, The British Library.

The sources of invention. J. Jewkes, D. Sawers and R. Stillerman. 2nd ed., 1969, Macmillan (really a textbook, but includes numerous case studies on the development of inventions).

Other books on invention are normally classified at 608 or 609 in the Dewey sequence in public libraries.

Patent information on the Internet

The Internet holds a great deal of information but this is often very difficult to identify. Old, as opposed to modern, patent information is scattered and less copious. Some useful sites are 'About.Com: inventors' at **http://inventors.about.com/ education/inventors/**; 'Inventure place' at **http://www.invent.org**; and the 'Jerome and Dorothy Lemelson Center for the Study of Invention and Innovation' at **http://www.si.edu/lemelson**. Otherwise it is worthwhile using the Altavista web site, **http://www.altavista.com** by running a search such as <+ 'thomas edison' +patents> in the search box. Many people are surprised how quickly very relevant material can often be found by such a strategy. It will often also pick up URLs of companies talking about their past successes.

For modern patents, an excellent starting point is the British Library's own site **http://www.bl.uk/patents**, which includes links to numerous databases (many free) of patent data, as well as to general information resources for inventors. The site is biased towards British sources but is a good starting point for anybody. The 'key databases' link heading lists important free databases which can be used to research what has been done before. These databases can generally be searched by keyword, classification, company and inventor name. Often Adobe Acrobat software is necessary to see images of the actual patent specifications.

However, these sources are of little use if you are unable to construct a good search, or to interpret the results. Nearly all of these databases are different in appearance from each other and it may be difficult to understand them. A good starting point is therefore to begin at a patent library. Both Britain and the USA have networks of publicly available patent collections. These are:

Britain's Patents Information Network (see **http://www.bl.uk/services/stb/ pinmenu.html** or ring the British Library on 020 7412 7919/20 for details of a nearby library).

The United States' Patent and Trademark Depository Library program (see **http://www.uspto.gov/go/ptdl**).

Those living in other countries can try the list of patent offices at **http://www.bl.uk/services/stb/etalmenu.html#indu**.

Contacts

The British Library itself holds over 44 million patent specifications from 38 countries. It is unique in that the national patent collection is not in the Patent Office itself, but rather in the national library, as this was thought to be useful for consultation of the science and technology collections.

Finally, we here at the British Library can offer *limited* free help to enquirers on both old and new patents. We cannot provide a free search to see if an idea has been done before (but will advise visitors on how to begin a search). We can often help with specific, known inventions, especially if they are historical in nature. In other cases we can often provide (non-legal) advice or make a suggestion. Please be as specific and detailed as possible. We can be contacted by phone (UK callers only) on 020 7412 7919/20, by fax on 020 7412 7480, by e-mail on patents-information@bl.uk or by post at Patents Information, The British Library, 96 Euston Road, London NW1 2DB. A great deal of information about what we do, what we hold and what services we offer, including copies of specifications, can be found through our web site **http://www.bl.uk/ services/stb/patents.html**. For legal enquiries it is best to contact the Patent Office on 0645 500 505 (UK callers) or by e-mail on patent-enquiries@ patents.gov.uk. The American Patent Office is not open to e-mail enquiries and should be contacted by phone on 800 786 9199.

INDEX

This index covers the 100 inventions, the names of the inventors and companies, and their countries of origin or residence. For the USA the states themselves are indexed, and for the UK the counties.